T0250506

Circuit Design Considerations for Implantable Devices

RIVER PUBLISHERS SERIES IN ELECTRONIC MATERIALS AND DEVICES

Series Editors

Edoardo Charbon
EPFL
Switzerland

Mikael Östling
KTH Stockholm
Sweden

Albert Wang
University of California
Riverside, USA

Indexing: All books published in this series are submitted to the Web of Science Book Citation Index (BkCI), CrossRef and to Google Scholar.

The "River Publishers Series in Electronic Materials and Devices" is a series of comprehensive academic and professional books which focus on the theory and applications of advanced electronic materials and devices. The series focuses on topics ranging from the theory, modeling, devices, performance and reliability of electron and ion integrated circuit devices and interconnects, insulators, metals, organic materials, micro-plasmas, semiconductors, quantum-effect structures, vacuum devices, and emerging materials. Applications of devices in biomedical electronics, computation, communications, displays, MEMS, imaging, micro-actuators, nanoelectronics, optoelectronics, photovoltaics, power ICs and micro-sensors are also covered.

Books published in the series include research monographs, edited volumes, handbooks and textbooks. The books provide professionals, researchers, educators, and advanced students in the field with an invaluable insight into the latest research and developments.

Topics covered in the series include, but are by no means restricted to the following:

- Integrated circuit devices
- Interconnects
- Insulators
- Organic materials
- Semiconductors
- Quantum-effect structures
- Vacuum devices
- Biomedical electronics
- Displays and imaging
- MEMS
- Sensors and actuators
- Nanoelectronics
- Optoelectronics
- Photovoltaics
- Power ICs

For a list of other books in this series, visit www.riverpublishers.com

Circuit Design Considerations for Implantable Devices

Editor

Peng Cong

Verily Life Science, US

River Publishers

Published, sold and distributed by:
River Publishers
Alsbjergvej 10
9260 Gistrup
Denmark

River Publishers
Lange Geer 44
2611 PW Delft
The Netherlands

Tel.: +45369953197
www.riverpublishers.com

ISBN: 978-87-93519-86-2 (Hardback)
   978-87-93519-85-5 (Ebook)

©2018 River Publishers

Contents

Preface

To my beloved wife, Rui

This book is based on a tutorial lecture, "Circuit Design Considerations for Implantable Devices," I gave at the IEEE International Solid-State Circuits (ISSCC) in San Francisco, CA, February 2016. Recently, there are increasing interests in implantable medical devices, especially implantable neuromodulation devices. Implantable devices are unique area for circuit designers. Comprehensive understanding of design trade-offs at system level is important to ensure device success. To expand the scope of the tutorial, many world-class experts in the field contribute much to this book, Dr. Kunal Paralikar (Medtronic Inc, US), Dr. You Zou (Verily Life Science, US), Dr. Zhou Wang (Princeton University, US), Dr. Naveen Verma (Princeton University, US), Dr. Abdollah Mirbozorgi (Georgia Institute of Technology, US), Dr. Maysam Ghovanloo (Georgia Institute of Technology, US), Dr. Ian Williams (Imperial College London, UK), Dr. Lieuwe Leene (Imperial College London, UK) and Dr. Timothy G. Constandinou (Imperial College London, UK). The goal of this book is to provide knowledge to circuit designers with limited biomedical background to understand design challenge and trade-offs for implantable devices. This book will also be welcomed by those readers with limited circuit design knowledge who need to understand the essentials of design challenge and trade-off on circuit design considerations.

Peng Cong

List of Contributors

Peng Cong, *Verily Life Science, US*

Kunal Paralikar, *Medtronic Inc., USA*

You Zou, *Verily Life Science, US*

Zhou Wang, *Princeton University, US*

Naveen Verma, *Princeton University, US*

Abdollah Mirbozorgi, *Georgia Institute of Technology, US*

Maysam Ghovanloo, *Georgia Institute of Technology, US*

Ian Williams, *Imperial College London, UK*

Lieuwe Leene, *Imperial College London, UK*

Timothy G. Constandinou, *Imperial College London, UK*

List of Figures

List of Tables

List of Abbreviations

ADC	Analogue to Digital Converter
ASIC	Application Specific Integrated Circuit
AVA	All-vs.-All
BMI	Brain Machine Interface
CDS	Correlated Double Sampling
CMOS	Complementary Metal-Oxide Semiconductor
CMRR	Common-Mode Rejection Ratio
CNS	Central Nervous System
DAC	Digital to Analogue Converter
DBS	Deep Brain Stimulation
ECG	Electrocardiography
ECOG	Electrocorticography
EEG	Electroencephalography
EKG	Electrocardiography
FDA	Food and Drug Administration
FSMs	Finite-state machines
IC	Integrated Circuit
IR	Infra-Red
LED	Light Emitting Diode
LFP	Local Field Potential
MEA	Micro Electrode Array
NEF	Noise Efficiency Factor
NGNI	Next Generation Neural Interfaces
OOK	On-Off Keying
PCB	Printed Circuit Board
PEDOT	Polyethylenedioxythiophene
PNS	Peripheral Nervous System
SAR	Specific Absorption Rate
SD	Secure Digital

SNR	Signal to Noise Ratio
SOC	System on Chip
SRAM	Static Random Access Memory
SVM	Support Vector Machine
VCO	Voltage Controlled Oscillator

1

Introduction

Peng Cong

Verily Life Science, US

1.1 Implantable Devices

Implantable medical devices, such as pacemakers, spinal cord stimulators, and cochlear implants, play increasingly important role in saving and improving the quality of people's lives. Circuit design is one of the important enabling technologies for those devices. It is helpful for the circuit designers to understand the uniqueness of implantable medical devices in order to make big contribution.

There are many commercially available implantable devices. A pacemaker, as shown in Figure 1.1(a), is a medical device to regulate heartbeat. It monitors electrocardiography (ECG or EKG) signal using sensing electronics and delivers an electrical stimulation through electrodes using stimulation circuit when missing heartbeat was detected. Due to accident or neurological diseases, some patients experience chronic pain with no explainable reasons and no conservative therapy works. In some cases, they can benefit from spinal cord stimulators as shown in Figure 1.1(b). A spinal cord stimulator is a device [1], which delivers pulsed electrical stimulation to the spinal cord to minimize chronic pain using similar stimulation circuit as a pacemaker, although the amplitude and stimulation frequency are much higher. Cochlear implants, as shown in Figure 1.1(c), may help provide hearing through stimulation of hearing nerve within cochlear in patients who are deaf because of damage to sensory hair cells.

In spite of different device functionalities, the electronic system of those devices has much in common. As shown in Figure 1.2, the system may include energy source with power management, data telemetry, sensing block, signal processing for algorithm, system control, and therapeutic stimulation sub-block. For disease intervention, the therapeutic stimulation is the most

1

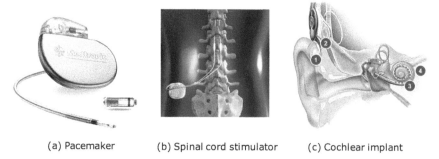

(a) Pacemaker   (b) Spinal cord stimulator   (c) Cochlear implant

Figure 1.1 Example of implantable medical devices: (a) a pacemaker, (b) spinal cord stimulator, and (c) cochlear implant.

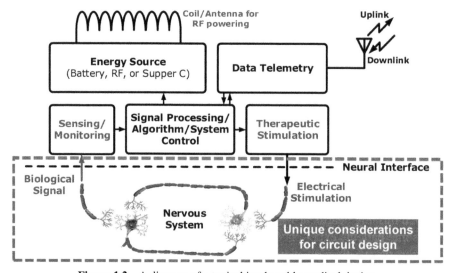

Figure 1.2 A diagram of a typical implantable medical device.

important sub-blocks, which interacts with nervous system through therapeutic electrical stimulation. Sensing is another key block, with signal processing unit, to determine disease state or abnormal nervous system behavior in order to delivery effective therapy, e.g., detecting a missing heartbeat for a pacemaker. Bi-directional neural interfacing with both sensing and stimulation capability is the key challenge and the uniqueness associated with implantable medical devices. As an isolated device implanted in human subjects, energy source is one critical sub-block in the system. The energy source could be a primary battery, a rechargeable battery, or continuous wireless powering coupled through electromagnetic field. The choice of the energy source highly depends on application desirability and technical feasibility. For example, primary battery is typically selected for pacemaker due to its low power consumption (stimulation frequency is typically smaller than 5 Hz and stimulation voltage is typically less than a couple volts) and highly reliability requirement. Rechargeable battery is typically selected for spinal cord stimulator due to it high-power consumption (stimulation frequency could be more than 100 Hz and stimulation voltage could be larger than 10 V).

1.2 Neural Interface

Understanding the challenges of the neural interface-associated circuit design considerations is critical for successful medical devices. Human nerve system was typically divided into central nervous system and peripheral nervous system. The central nervous includes brain and spinal cord. The brain is the center of the nervous system with more than 86 billion neurons (compared to 5.5B transistors in the latest commercially available CPU as of 2015). Spinal cord connects brain to peripheral nervous system, which then controls limbs and organs, and sends sensational signal back to brain or spinal cord. The nervous system is mainly formed by neurons. One basic function of neuron is to transfer information through both electrical and chemical activities. Figure 1.3 depicts two neurons communicating with each other. When an electrical signal, typically called action potential or spike, is generated, it flows down the axon of the cell on the left, with the help of ions such as sodium, potassium, and calcium to transmit the electrical message. When it reaches its terminal gap with the next neuron, a neurotransmitter, called synapse, will be released and passed along to the next cell. The neurotransmitter will help generate another spike in the next neuron. Using drug to regulate neurotransmitters is one way to do intervention; however, for an implantable device, it typically uses stimulation through electrodes [2].

If there is an ideal electrode, which can be very small, close to neuron membrane even inside the neuron, and have a zero impedance between electrode and the medium around, as shown in Figure 1.3, a spike can be measured with resting potential about -70 mV, active potential level roughly $+30$ to 40 mV and the pulse width ~ 1 ms. Action potentials are generated by voltage-gated ion channels in a cell's membrane. These channels are shut when the membrane potential is near the resting potential of the cell, but they rapidly begin to open if the membrane potential increases to a

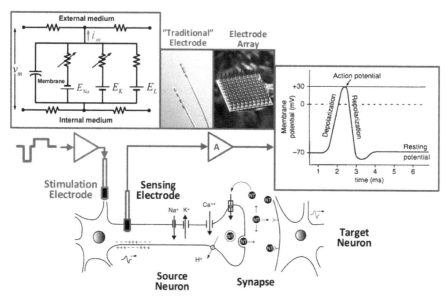

Figure 1.3   Examples of neuron communication and a bi-directional neural interface.

precisely defined threshold value. When the channels open, they allow an inward flow of sodium ions, which change the electrochemical gradient, and in turn produce a further rise in the membrane potential. The process proceeds explosively until all of the available ion channels are open, resulting in a large upswing in the membrane potential. The rapid influx of sodium ions causes the polarity of the plasma membrane to reverse, and the ion channels then rapidly inactivate. As the sodium channels close, sodium ions can no longer enter the neuron, and then they are actively transported back out of the plasma membrane. Potassium channels are then activated returning the electrochemical gradient to the resting state. These action potentials can be engaged by using therapeutic stimulation in neuromodulation.

Fundamentally, for neural interface, electrodes are used to listen to neural communication and to apply stimulation to control neural communication. Most of the readers, who are interested in this book, must know how to use an oscilloscope to measure a voltage on a PCB. Oscilloscope probes are typically used to interface with the test points on the PCB, where it makes a metal-to-metal ohm contact; therefore, what displayed on the scope matches the voltage at the test-point. It would be a different case for electrode–tissue interface, which would be discussed in details in Chapter 2.

1.3 Circuit Design Considerations for Neural Sensing

Typically, there are two different types of signal, which neural interface circuit works with, spike and local field potential (LFP). Spike is the fundamental signal within neural system. Some microelectrodes can pickup neural spike; however, many electrodes are not ideal in reality. They are not small enough to pick up signal from only one neuron, not close enough to pick full scale of the signal, and have finite impedance resulting in signal distortion. Therefore, in an implantable system, this signal is rarely seen. Instead, the sensed signal it typically much small in amplitude with slightly different characteristics. The signal bandwidth is typically in kHz range. The other typical signal is LFP, which is an electrophysiological signal generated by the summed electrical signal from 100 to 1000's of nearby neurons within a small volume of nervous tissue. The signal bandwidth is much reduced up to a few hundreds of Hertz due to the averaging effect and presentation of the signal is oscillation in nature instead of spike. However, this signal may contain more high-level biological meaning compared to spikes. Spikes can typically be measured by using micro wire electrode or microelectrodes array and LFP can be accessed by more "traditional" electrodes as shown in Figure 1.3, and the choice of spike or LPF strongly depends on applications and this can have a big impact on the architectural choices.

Electrodes are used to detect neural signals; however, electrodes are not ideal. When an electrode, which is usually metal, is placed inside a physiological medium (e.g., tissue or body fluid, which contain electrolytes), electrochemical activity will occur, and an interface will form between the two phases. In the metal electrode phase, charges are carried by electrons; in tissue (or, more generally, in the electrolyte phase), charges are carried by ions. At zero biasing, a layer of oriented water molecules with two layers of charges can be found in the interface. This behaves as a capacitor with a small signal and is called a double-layer capacitor (C_D), as illustrated in Figure 1.4.

There is also a resistive component associated with the electrolyte, R_S. The values of both C_D and R_S are a function of electrode geometry, material, and surface profile. As the size decreases from traditional electrodes to microelectrodes, the value and mismatch tend to increase. As shown in Figure 1.5, with the input impedance of the amplifier R_{In}, the two electrodes can be replaced by the electrode model R_S and C_D. At the same time, the neural circuit can be replaced by an ideal signal source. A few circuit design considerations can be determined by the circuit representation. The first is the mismatch between electrodes (R_{S1}, C_{D1} and R_{S2}, C_{D2}), and the second is the

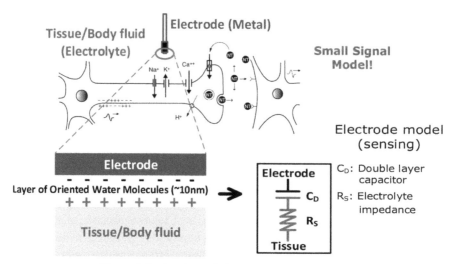

Figure 1.4   An electrode interface model for sensing.

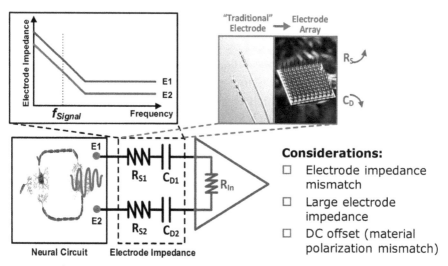

Figure 1.5   Circuit design considerations for a sensing neural interface.

high electrode impedance, especially when electrodes move from traditional ones to microelectrodes. A large mismatch results in high signal susceptibility to common-mode interference, and large electrode impedance results in large signal attenuation and, as a consequence, a more pronounced mismatch.

The electrode performances, for both matching and impedance values, are typically beyond the control of circuit designers. Generally, improving input impedance and common-mode rejection ratio (CMRR) helps for both cases. Another important consideration for electrodes is polarization, which results in a DC offset between two electrodes. This polarization could be caused by material and tissue mismatch or stimulation effect. The DC offset can typically be managed by using an AC-coupled amplifier design.

Based on the nature of neural signal characteristics and the electrode model, a few circuit design considerations can be extrapolated:

- DC offset: an AC-coupled or AC-coupled equivalent (e.g., with DC servo to remove DC offset) amplifier
- Electrode impedance mismatch: large circuit input impedance and high CMRR
- Large electrode impedance, especially for microelectrodes: large circuit input impedance.

In addition, the following factors should be considered. For spike detection, there is generally a need for low-power, low-noise (i.e., thermal noise), small size, and high-input impedance. As the signal band is typically 300 Hz to 10 kHz, $1/f$ noise may not be very critical. However, high-input impedance and small size are critical, as these are often required when working with microelectrodes and large electrode counts. For LFP, requirements typically include low power, low noise (i.e., thermal noise and $1/f$ noise), and high-input impedance. As the bandwidth of LPF is from sub-Hz to 300 Hz, minimizing $1/f$ noise is important. The impedance requirement is not as rigid as for the spike amplifier, but a reasonable value will help system performance.

Circuit design considerations and circuit architecture for different sensing amplifiers would be presented in Chapter 3.

1.4 Circuit Design Considerations for Stimulation

Typically, two electrodes are needed for stimulation, where one as reference and the other as stimulation electrode. The goal of stimulation is to delivery charge into tissue to influence the potential outside of neuron membrane so that to initiate spikes/action potential. A typical stimulation waveform is shown in Figure 1.6 with the x-axis time and the y-axis current. The positive and negative indicate the direction of the charge flow, in or out of tissue.

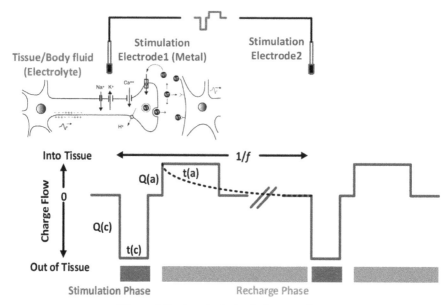

Figure 1.6 A typical stimulation waveform.

The waveform typically includes stimulation phase followed by recovery phase. It could be an active recharge by using the same or scaled current with reversed direction, or passive recovery by shorting the electrodes as shown in the dash-line, which follows RC discharge curve.

In the case of sensing, the electrode and tissue interface can be considered as a pure capacitor, where there is no charge going across the double layer structure. However, in the case of stimulation (typical stimulation voltage could vary from 1 to 10 V), there could be real charge transfer with electrochemical reaction. The leakage through the double layer capacitor is typically called Faradeic impedance. The model of the electrode interface for stimulation is shown in Figure 1.7. Note that the value of the Faradeic impedance may vary with stimulation amplitude and other factors.

The most important consideration for stimulation circuit design is safety, which largely depends on whether the electoral-chemical reactions are reversible or irreversible. If the products do not diffuse away during the stimulation phase, it could be recovered during discharge phase. In this case, although electron transfer occurs, in terms of the electrical model, it is better modeled as a capacitor. So it was named pseudocapacitance, one example is Pt. During the stimulation phase, the reaction product of electron, hydrogen

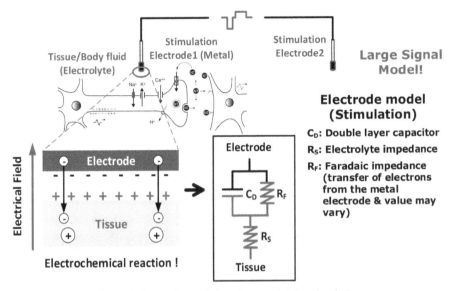

Figure 1.7   An electrode interface model for stimulation.

ions, and platinum is Pt–hydrogen bond, and these products only stay at the electrode surface (called hydrogen plating) thanks to the characteristic of Pt. If the total charge injected into tissue during stimulation phase is totally recovered during recharge phase (charge-balanced stimulation), all the production will be recovered too. This is the reason why Pt is used widely in implantable devices. In short, for a reversible case, charge-balanced stimulation is critical to ensure the safety.

If the products diffuse away from the electrode, the charge cannot be recovered upon reversing the direction of current. One example is water reduction, where hydrogen will be formed. Therefore, control voltage cross double layer capacitor within a limitation is essential in this case with the control of total DC current under a limit.

Beside the electrode safety considerations, neuron safety is an even bigger topic and so far people still do not have a complete understanding to determine neuron safety under stimulation. A wide accepted standard is called Shannon criteria illustrated in Figure 1.8, proposed by professor Shannon in 1992. Note the criteria is pure experimental by plotting previous results against charge per phase (*x*-axis) and charge densities (the charge divided by the area of electrode) per phase (*y*-axis). The line in Figure 1.8 is defined by the equation below:

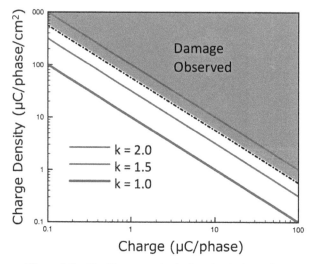

Figure 1.8   The Shannon criteria for stimulation safety.

$$k = \log(Q) + \log(Q/A),$$

where, Q is charge per phase and A is the area of electrode. If K is greater than 1.5–1.8, it would not be safe. Shannon criteria have been evolved in recent years for microelectrodes.

Below is the summary of circuit considerations for stimulation:

1. Work with electrode designer and understand electrode model (careful! Low signal vs. large signal model).
2. Small DC leakage current from electrode to tissue (<1 μA or 10 nA).
3. Charge balanced stimulation for reversible reaction.
4. Monitor and control voltage across double layer capacitor within "good" reaction window if irreversible reaction might happen.
5. Limit maximum charge and charge density for neuron safety.
6. Constant current versus constant voltage stimulation.
7. Power efficient design (Stimulation therapy circuit typically is the major power consumption source in implantable devices).
8. "Good" performance (frequency, stimulation amplitude, and stimulation pulse width accuracy).
9. Pay more attention to micro-electrodes!

The detailed safety consideration and circuit design of stimulation would be presented in Chapter 2 and Chapter 4, respectively.

1.5 Closed-Loop System

As shown in Figure 1.2, the other important module of the system architecture is system control unit processing unit, which often runs signal processing and algorithm control. One example of how the module involved in the system can be found in cardiac devices, e.g., pacemakers. Pacemakers monitor ECG or EKG signal using sensing electronics. The EKG signal is analyzed by on-device algorithm unit. An electrical stimulation through electrodes is then applied if a missing heartbeat was detected. As captured in Figure 1.9, the pathway to classification generally involves a series of steps: from raw biosignal recording to initial feature extraction, e.g., identifying the peak of an EKG; to detection of some feature of relevance, e.g., the peak of the R-wave and the rhythm associated with ventricular contraction in the heart; finally to the classification and to the estimation of the potential state of the patient, e.g., atrial fibrillation if the spacing between R-wave feature peaks becomes erratic and/or extraordinarily fast. The process from signal extraction to classification requires knowledge of the desired information that is relevant for the clinician and/or an algorithm, as well as a method of signal processing for deriving that classification. Identifying the desired classification is often collaborated with biomedical engineers; but the signal processing is domain knowledge relevant to good circuit design.

The classification of EKG can be achieved by using time domain analysis with "small" power consumption and "low" computation capability. One

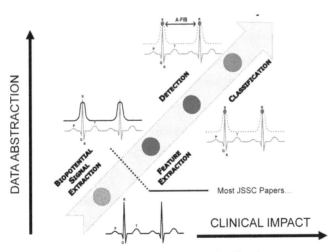

Figure 1.9   Series of steps of classification.

example of system architecture can be found in Figure 1.10 (a). However, when researches move into central nerve system, e.g., brain, the analysis gets more involved, which is sometimes involved in frequency domain analysis as illustrated in Figure 1.10(b). In addition, instead of single channel analysis as in pacemaker, brain interface device need multiple channels up to hundreds to thousands, which post-significant challenge to embedded system design with significant power constrain within implantable devices. Chapter 5 will present latest research in this area.

1.6 RF Powering and Data Telemetry for Implantable Devices

One of the most difficult challenges for implanted device technologies to overcome is in providing implants with sufficient power. There has been great progress in understanding how to design miniature low-power circuits for biological applications, which enables devices, such as pacemakers, via primary batteries with reasonable size. However, batteries are difficult to miniaturize and remain the size-limiting component of many implants. In addition, the lifetime of batteries limits the useful life of potential implants.

Battery replacement for implantable devices often requires an additional surgery and can cause many complications. Alternatively, rechargeable batteries allow for longer useful lifetimes but need an additional means of delivering power to recharge, such as RF approaches, which suffer from low-efficiency power transfer and require relatively large, aligned antennas. Other non-RF methods to wirelessly power implanted devices have been proposed but are only in very early stages of development and will require many advances before they are practical. WiTricity, which uses magnetic resonance coupling, allows for highly efficient energy transfer but requires large coils [3]. Ultrasound energy can be used to deliver power to implanted devices, but the efficiency of power delivery is very small [4]. Energy scavenging [5] and optical energy [6] have also started to be investigated but currently produce too little energy to reliably power implantable devices.

Currently, wireless powering through inductive link is still the main method for implant with rechargeable battery. A typical model of such a system can be found in Figure 1.11(a). L_1 is the coil of external charger and L_2 is the coil of implant. The mutual inductance of the two coils services the link of powering. An analysis of this configuration, where two circuits are coupled together through a mutual inductance, is inconvenient to perform

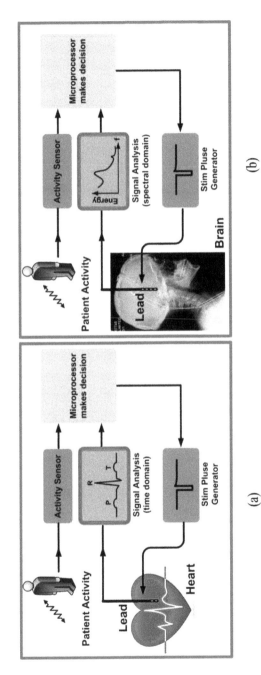

Figure 1.10 Closed-loop architecture for (a) pacemakers and (b) brain interface devices.

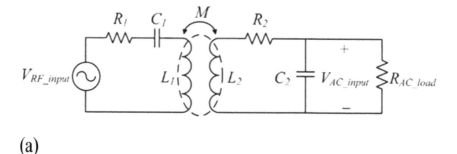

(a)

(b)

Figure 1.11   Equivalent circuit of wireless powering with inductive link.

calculation and difficult to understand. Fundamental circuit theory reveals that the coupled circuit can be separated of de-coupled by employing a reflected impedance at the primary circuit, as illustrated in Figure 1.11(b)

$$Z_{\text{ref}} = \frac{(\omega M)^2}{Z_{\text{s}}},$$

where Z_{S} is the total impedance of the secondary circuit without the primary circuit loading. In addition, an induced voltage,

$$V_{\text{ind}} = -j\omega M I_{\text{p}}$$

where I_{p} is the primary circuit current. Many system designs and design trade-off can be derived from this model.

For many implantable devices in the market, metal (typically titanium or titanium alloys) is used as device case to ensure hermeticity. As wireless powering coil is typically located inside the metal can, low frequency (less than 1 MHz) with large skin depth is chosen to ensure efficient energy transferred into the device. For some devices, such as cochlear implant, the powering coil is located outside device case, higher powering frequency is typically used for small device/coil size. However, it is typically limited to a few tens of MHz. Recently, GHz wireless powering has been studied. It is typically limited to low-power application with average power consumption of a few microwatts. The design approach is typically to optimize wireless powering efficiency while maximizing external charger power. Typically RF powering at low frequency (e.g., <10 MHz) is limited by heat generation and RF powering at high frequency (e.g., GHz) is limited by specific absorption rate (SAR). Received power may vary much due to position/orientation induced coupling coefficient variation Power management/protection. Advancing in this field will be presented in Chapter 6.

References

[1] Windhover Information Inc (2007). *Neurostimulation Market Expanding*. Norwalk, CT: Windhover Information Inc.

[2] Durand, D. B. (2006). "Electrical stimulation of excitable systems," in *Biomedical Engineering Fundamentals*, ed. J. D. Bronzino (Boca Raton: CRC Press).

[3] Kurs, A., Karalis, A., Moffatt, R., Joannopoulos, J. D., Fisher, P., and Soljacic, M. (2007). Wireless power transfer via strongly coupled magnetic resonances. *Science* 317, 83–86.

[4] Lee, K. L., Lau, C. P., Tse, H. F., et al. (2007). First human demonstration of cardiac stimulation with transcutaneous ultrasound energy delivery: implications for wireless pacing with implantable devices. *J. Am. Coll. Cardiol.* 50, 877–883.

[5] Justin, G. A., Zhang, Y., Sun, M., and Sclabassi, R. (2004). "Biofuel cells: a possible power source for implantable electronic devices," in *Proceedings of the 26th Annual International Conference of the IEEE Engineering in Medicine and Biology Society*. California, CA.

[6] Murakawa, K., Kobayashi, M., Nakamura, O., Kawata, S. (1999). A wireless near-infrared energy system for medical implants. *IEEE Eng. Med. Biol.* 18, 70–72.

2

Neural Stimulation and Interface

Kunal Paralikar

Medtronic Inc., USA

Implantable neurostimulators interface with the nervous system and deliver energy to activate it through application specific neural interfaces, also known as electrodes. In this chapter, a grounds-up approach is adopted to allow the reader to appreciate the three elements that participate in neural stimulation and sensing, namely the neuronal element, the tissue element, and the tissue–electrode element.

2.1 Neuronal Element

Research into the causes of neurological diseases affecting motor functions, cognition, and emotional well-being increasingly points toward malfunctions in neural circuits [1]. Such circuit malfunctions can impact the brain's computational and information flow capabilities, leading to clinical symptoms that negatively impact bodily functions and quality of life. Neurons are the fundamental building blocks of the nervous system and transmit information via both electrical and chemical signals. Electrical signals are typically used to transmit information for long distances within neurons while chemical signals are used to communicate between the neurons at short distances.

2.1.1 Neuron Structure

Structurally, a neuron can be divided into three segments, namely: (i) cell body, (ii) dendrites, and (iii) axon (Figure 2.1). These segments together are responsible for the neuron's essential function of receiving incoming information, processing that information, and communicating information to targets such as other neurons, glands, or muscles.

17

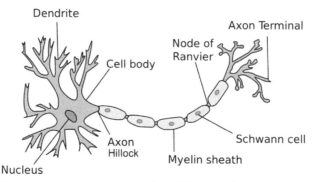

Figure 2.1   Neuron. Inputs are received at the cell body and dendrites. Action potentials initiate at the axon hillock and terminate at the axon terminal. Their propagation down the axon is aided by insulating myelin sheath secreted by the Schwann cells. Nodes of Ranvier enable regeneration of the action potential during its propagation.

Source: https://commons.wikimedia.org/wiki/File:Neuron.svg

The cell body or soma of the neuron is similar in function to other cells; it is responsible for the cells basic function, contains the nucleus, and produces neuronal proteins. Together with the cell body, the dendrites receive and process incoming information. Neurons typically receive numerous input signals (hundreds to thousands and more depending on the type of neuron) through its dendritic tree. These input signals can be excitatory or inhibitory in nature and are graded (analog levels).

The cell body and dendrites process these signals to determine if the overall response of the neuron is excitatory or inhibitory. Excitatory response will induce a neuron to trigger a nerve impulse or an action potential (electrical signals), whereas inhibitory response will prevent it. The nerve impulses originate at the axon hillock which is at the junction between soma and the axon.

Axon is typically longer than dendrites and functions to transmit the nerve impulse to target cells. It is covered with myelin sheath that acts to insulate the axon and allows for propagation of the electrical signals in a regenerative manner over longer distances without loss of strength. Uninsulated gaps in the myelin sheath called the Nodes of Ranvier contain protein channels embedded in the cell membrane which enable the exchange of ions with the extracellular space, a process which sustains and regenerates the propagating action potential. Since there is little loss to the strength of the action potential as it travels down the axon, neurons encode information through both rate (number of action potentials per unit time) and temporal pattern (time between action potentials).

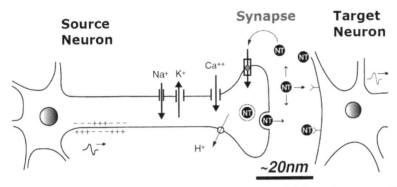

Figure 2.2 Communication of electrical impulse/action potential down the axon results in release of neurotransmitters in the synapse which are absorbed by the receptors located on dendrites and cell bodies of down-stream neurons.

Action potential propagation terminates at distal end of the axon and the electrical signals are converted to chemical signals in the form of neurotransmitters. These chemicals are secreted into the gap space between the axon of one neuron and the dendrite of another neuron called the synapse, in a graded manner thats dependent on the volume and frequency of incoming action potentials in the axon. Whether a given neurotransmitter will have an excitatory or inhibitory effect on the downstream neuron will depend only on the type of receptors the downstream neuron possesses (Figure 2.2).

The axon is considered to be the most important structural unit of the neuronal element within the electrical stimulation framework (the most common form of neural stimulation) because it is the easiest to activate [2]. Action potentials sensed in the extracellular space are typically recorded from cell-bodies in case of an unmyelinated neuron and from axons in case of a myelinated neuron.

2.1.2 Neuron: Electrical Characteristics

In the description that follows, we will evaluate the electrical characteristics of the neuronal element in steady-state (in absence of external electrical stimulation), derive equation for potential across the cell-membrane, understand the dynamics of the action-potential, derive equation to describe electrical behavior of an axon, and then introduce the concept of extracellular stimulation and how it impacts all of the above.

Biologically, an axon can be considered as a banana immersed in sea water, i.e., it has an excess of potassium ions inside it and an excess of sodium ions outside it. Electrically, an axon can be thought of as a conducting wire covered with insulation that has selective permeability to different types of ions along its length. As a result, current flow in axon is both parallel and perpendicular to its length.

2.1.2.1 Membrane model and resting membrane potential

Perpendicular current flow (I_m) is best described by the Hodgkin and Huxley model of the neuronal membrane (refer to Figure 2.3) The variable resistor battery segments for different ions [Sodium (Na$^+$), Potassium (K$^+$), Chlorine (Cl$^-$)] depict the ion channels that are membrane-spanning proteins. Their permittivity to the ions changes in response the membrane voltage (V_m). The membrane voltage is the potential difference between the inside and outside of the cell ($V_m = V_{in} - V_{out}$).

From Figure 2.3, the membrane current $I_m = I_c + I_I$, where I_c is the capacitive current $= C_m \frac{dV_m}{dt}$, and I_I is the sum of ionic currents due to sodium, potassium, chlorine ions, and leakage $= I_{Na} + I_K + I_{Cl} + I_{leak}$. Current due to Cl$^-$ ions is typically negligible.

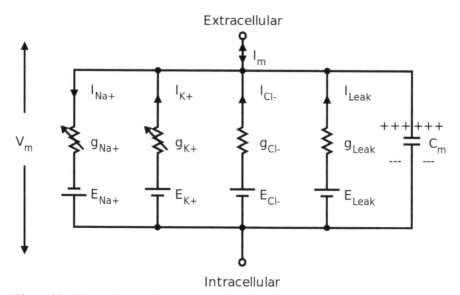

Figure 2.3   Electrical equivalent representation of a cell membrane. g_x represents variable conductance to ions X and leakage currents, E_x represents their Nernst Potential and C_m is the membrane capacitance.

Source: https://commons.wikimedia.org/wiki/File:MembraneCircuit.svg

The individual ionic currents are expressed as:

$$I_{Na} = g_{Na}(V_m - E_{Na});$$
$$I_K = g_K(V_m - E_K);$$
$$I_{Cl} = g_{Cl}(V_m - E_{Cl});$$
$$I_{Leak} = g_{Leak}(V_m - E_{Leak}).$$

The terms: E_{Na}, E_K, E_{Cl}, and E_{Leak} in the above equation are called the Nernst Potentials. Two types of gradients contribute to ionic flow; the concentration gradient and the potential gradient. The difference in concentration (concentration gradient) across the membrane causes movement of charge across it called the diffusion current. This movement of charge in turn creates a potential gradient (an electric field) across the membrane resulting in conduction currents. The potential at which the conduction and the diffusion currents are in equilibrium resulting in net zero ionic current is called the Nernst Potential. It is given by

$$E_{ion} = \frac{RT}{zF} \ln \frac{[ion]_{out}}{[ion]_{in}},$$

where

R is the universal gas constant (8.31 J/mol K)
T is the temperature in $°K$
z is the valence of the ion
F is the Faraday's constant (96,485 C/mol)
$[ion]_{out}$ is the extracellular ionic concentration
$[ion]_{in}$ is the intracellular ionic concentration.

Each ion type will have a unique Nernst Potential for each type of neuron. For example, for a giant squid axon, the Nernst Potential for sodium and potassium is 50 and –77 mV, respectively.

The neuronal membrane is a highly dynamic environment. Even in the absence of the any stimulus; ion pumps that exchange $2Na^+$ ions for $3K^+$, ionic gradients, selective, and non-linear permeability of ion channels (described next) result in non-zero membrane potential called the resting membrane potential. Based on the research conducted by scientists Hodgkin, Katz, and Goldman [3, 4], it was found that the resting membrane potential depends on temperature, intracellular, and extracellular concentrations of individual ions, valence, and selective permeability of ions. Specifically, for neuronal membrane with Na$^+$, K$^+$, and Cl$^-$ ions, the resting membrane potential E_r is

$$E_r = -\frac{RT}{F} \ln \frac{P_{K^+}[K^+]_{in} + P_{Na^+}[Na^+]_{in} + P_{Cl^-}[Cl^-]_{out}}{P_{K^+}[K^+]_{out} + P_{Na^+}[Na^+]_{out} + P_{Cl^-}[Cl^-]_{in}}.$$

Note that in steady state, the resting membrane potential is equivalent to the transmembrane potential. For a typical neuron, in a steady-state, the resting membrane potential is approximately –70 mV.

2.1.2.2 Action potential

Now, let's consider how the membrane condition change when the transmembrane potential changes. When received by the dendritic receptors, excitatory neurotransmitters open various types of ion channels that depolarize (reduce the transmembrane potential) the cell. Typically, when the transmembrane potential reduces from –70 to –50 mV, and when this wave of depolarization reaches the axon hillock (if the depolarization caused by the neurotransmitters is strong enough to reach the axon hillock), it activates the voltage-gated sodium and potassium channels. Activation (conductivity) of the sodium-ion channels increases rapidly compared to that of the potassium-ion channels. As a result, sodium ions from the extracellular space rush into the cell and further depolarize it (i.e., making it more positive from its initial steady-state, raising the transmembrane potential). When the transmembrane potential increases to about 20 mV (from the initial –70 mV), the conductivity of the potassium-ion channels increases resulting in an outward flow of potassium ions. At the same time, the conductivity of the sodium-ion channels decreases. The net result is a decrease in the transmembrane potential and eventually the membrane reaches equilibrium voltage. The above description was that of an action-potential/nerve impulse at an axon cross section. Figure 2.4 shows an action potential recorded from an extra-cellularly placed microelectrode. The initial depolarization phase marked by an in-rush of sodium ions into the neuron is observed as a potential drop in the extracellular space. Similarly, the subsequent outflow of potassium ions from the neuron during is observed as a potential increase in the extracellular space.

The action potential is initiated at the Axon Hillock when the membrane potential due to depolarization is above a critical value, typically around -45mV. Action Potential is an all-or-none phenomenon, which means once it is initiated at the axon-hillock it propagates along the axon in a regenerative manner without loss of amplitude. This is because the voltage-gated ion

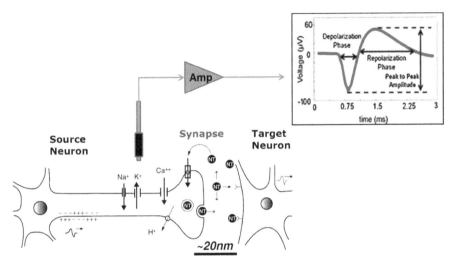

Figure 2.4 Schematic for recording extracellular action potential.

Source: Courtesy Dr. Cong and adapted from Clement et al. [6].

channels have an all-or-none (digital) response. It is important to note that typically the action-potential only travels in one direction from the axon-hillock to the axon terminal (presynaptic junction). Although in theory, an action potential that initiates in the center of the axon (and not at the hillock) can propagate in both directions. The dynamics of the voltage-gated channels is responsible for this behavior of the action potentials.

Specifically, the voltage-gated channels have three states: deactivated, activated, and inactivated. The channel cycles through these states in this order and spends most of its time in the deactivated state. The action potential occurs during the activated state, which is followed by the inactivated state (known as the refractory period), during which the ion channel cannot respond to changes in transmembrane voltage. As a result, once an action potential has passed through a given location of the axon, that location is refractory, prevents another action potential at that location, thereby giving a forward directionality to the action potential. These channel and transmembrane dynamics are best described mathematically by the Hodgkin–Huxley model [5]. It is summarized next "g_x" are channel conductances.

$$g_K = G_K n^4 \qquad g_{Na} = G_{Na} m^3 h \qquad g_{Cl} = G_L.$$

Where *n*, *m*, and *h* are given by

$$\frac{dm}{dt} = \alpha_m \left(1 - m\right) - \beta_m m \qquad \frac{dh}{dt} = \alpha_h \left(1 - h\right) - \beta_h h$$

$$\frac{dn}{dt} = \alpha_n \left(1 - n\right) - \beta_n n.$$

Where constant α_m, β_m, α_h, β_h, α_n, and β_n are empirically derived,

$$\alpha_m = \frac{0.1 * (25 - V')}{e^{\frac{25 - V'}{10}} - 1} \qquad \beta_m = \frac{4}{e^{\frac{V'}{18}}}$$

$$\alpha_h = \frac{0.07}{e^{\frac{V'}{20}}} \qquad \beta_h = \frac{1}{e^{\frac{30 - V'}{10}} + 1}$$

$$\alpha_n = \frac{0.01 * (10 - V')}{e^{\frac{10 - V'}{10}} - 1} \qquad \beta_n = \frac{0.125}{e^{\frac{V'}{80}}}$$

Where G_{Na}, G_K, and G_L are constants and $V' = V_m - E_r$.

The above description can be considered a single-compartment description wherein dynamics of current flow across the membrane is presented. The associated potential change inside the neuron causes a potential gradient and corresponding current flow inside the neuron, parallel to its length. This current flow and how it is modeled will be discussed in the next section.

2.1.3 Neuron: Cable Equation

Consider a small axonal segment of length Δx, with potential of V_n and $V_{(n+1)}$ at two ends and a resistance $R_i = r_i \times \Delta x$ in between them. Here, r_i is resistance per unit length (Figure 2.5).

Therefore,

$$i_{i(n+1)} = \frac{\left(V_{in} - V_{i(n+1)}\right)}{r_i \Delta x} = -\frac{1}{r_i} \frac{\Delta V}{\Delta x}$$

As $\Delta x - \to 0$

$$i_i = -\frac{1}{r_i} \frac{\partial V}{\partial x}.$$

Therefore, axial current flow inside the neuron per unit length is

$$\frac{i_i}{\Delta x} = -\frac{1}{r_i} \frac{\partial^2 V}{\partial x^2}.$$

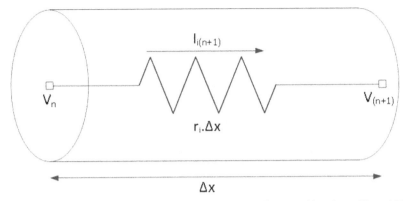

Figure 2.5 Intracellular current flow between nodes n and $n + 1$ with voltage V_n and V_{n+1}, respectively, and resistance per unit length of r_i.

It is proportional to the second spatial derivative of the voltage.

A similar approach can be adopted to extracellular current flow in response to extracellular voltage application. Note that in practical stimulation settings, as a result of constant current or constant voltage flow, an extracellular potential (V_e) gradient is created. This would in turn result in extracellular current flow.

$$i_e = -\frac{1}{r_e}\frac{\partial V_e}{\partial x}.$$

r_e is extracellular resistance per unit length. Therefore, current flow per unit length due to extracellular voltage is

$$\frac{i_e}{\Delta x} = -\frac{1}{r_e}\frac{\partial^2 V_e}{\partial x^2}.$$

Now consider current flow through the membrane due to both internal and external current flow. Assuming directions of current flow as shown in the Figure 2.6 and applying Kirchhoff' current law at node 1

$$i_{i(n+1)} - i_{i(n-1)} + i_m \Delta x + i_{e(n+1)} - i_{e(n-1)} = 0$$

where i_m is the membrane current per unit length

$$i_m = \frac{i_{i(n-1)} - i_{i(n+1)}}{\Delta x} + \frac{i_{e(n-1)} - i_{e(n+1)}}{\Delta x}.$$

Now,

$$\frac{i_{e(n-1)} - i_{e(n+1)}}{\Delta x} = -\frac{\partial i_e}{\partial x} = \frac{1}{r_e}\frac{\partial^2 V_e}{\partial x^2}.$$

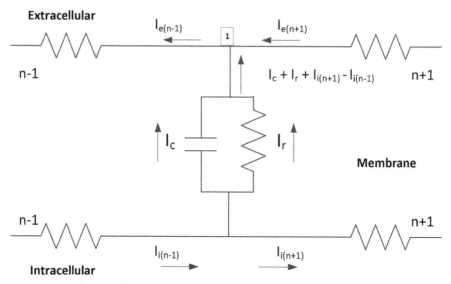

Figure 2.6   Ladder model of a neuron in unmyelinated location.

and

$$\frac{i_{i(n-1)} - i_{i(n+1)}}{\Delta x} = -\frac{\partial i_i}{\partial x} = \frac{1}{r_i}\frac{\partial^2 V}{\partial x^2}.$$

Note that V is potential inside the neuron and V_e is the extracellular potential. Therefore,

$$i_m = \frac{1}{r_e}\frac{\partial^2 V_e}{\partial x^2} + \frac{1}{r_i}\frac{\partial^2 V}{\partial x^2}.$$

Now,

$$i_m = i_c + i_r$$

$$= C_m\frac{\partial V}{\partial t} + \frac{V}{r_m}$$

Combining the above two equations,

$$C_m\frac{\partial V}{\partial t} + \frac{V}{r_m} = \frac{1}{r_e}\frac{\partial^2 V_e}{\partial x^2} + \frac{1}{r_i}\frac{\partial^2 V}{\partial x^2}$$

$$\lambda^2\frac{\partial^2 V}{\partial x^2} - \tau_m\frac{\partial V}{\partial t} - V = -\lambda^2\frac{\partial^2 V_e}{\partial x^2}$$

In this equation $r_e \approx r_i = r$ (assumed), $\lambda = \sqrt{\frac{r_m}{r}}$, $\tau_m = C_m \times r_m$.

This equation explains the currents generated due to extracellular electrical stimulation (V_e). It also quantifies voltage inside the neuronal membrane (V) in both space and time. However, there are numerous simplifying assumptions. (i) The membrane resistance (r_m) is dynamic, with dependence on transmembrane voltage as explained by the Hodgkin–Huxley model. (ii) The treatment here is similar for infinite length neurons whereas neurons have finite length. (iii) Importantly, neurons are branched and not cables. In spite of these limitations/assumptions, these equations give insight into the membrane dynamics.

In practice, neurons exhibit a wide variety in their morphology, fiber diameters, transmembrane channel type, density, and dynamics. All these factors together with electrode placement play a role in transmembrane potential and the associated neural response. To account for these factors, models with progressively higher details to simulate neural response to stimulation are incorporated in software packages like NEURON which are used extensively.

2.2 Tissue Element

Neural tissue is anisotropic and inhomogeneous with non-linear time, frequency, and amplitude dependent response to electrical stimulation. These properties of the tissue play an important role in determining the electric field around the neurons due to electrical stimulation.

Electric field due to charge density ρ, in medium with permittivity ϵ (Gauss' law)

$$\nabla \overline{E} = \frac{\rho}{\epsilon}.$$

Assuming quasi-static conditions, the curl of $\overline{E}$ is zero based on Faraday's law of induction, the relation between potential Φ and electric field is

$$\overline{E} = -\nabla \Phi.$$

Combining the above two equations

$$-\nabla^2 \Phi = \frac{\rho}{\epsilon}.$$

In tissue, the charge density is negligible, almost zero. As a result,

$$-\nabla^2 \Phi = 0.$$

The above equation is a Laplace equation and the goal is to solve this equation adhering to conditions set by the tissue and the stimulation electrodes. Finite

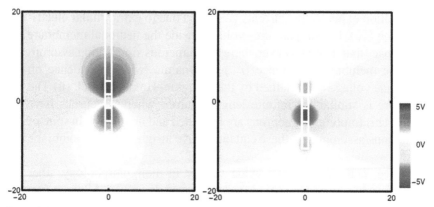

Figure 2.7   Potential distribution due to bipolar (a) and tripolar (b) stimulation configurations for electrodes used in spinal cord stimulation.

Source: Adapted from [7].

element analysis (FEA) is used to numerically solve the Laplace equation and to determine the potential distribution resulting from the electric field. The FEA analytic approach is used because the voltage field for practical electrodes is complicated. In Figure 2.7, simulation results showing the nature of potential distribution even when tissue is assumed to be isotropic homogenous medium are presented. The computations increase in complexity when detail such as tissue anisotropy due to tissue response, different tissue types, alignment of electrodes and neurons is incorporated to accurately predict the volume of neural activation.

A detailed discussion of use of modeling to analyze potential distribution in the tissue and corresponding neuronal excitation is outside the scope of this chapter. Interested readers are encouraged to read reference [8] to initiate their detailed study on this matter.

One of the important characteristic of the tissue is its impedance which will be covered next while describing the electrode–tissue interface.

2.3 Electrode–Tissue Element

Electrodes are the elements by which an engineered system (neurostimulator) interfaces with the biological system (neural tissue). It is essential to understand how the electrode interfaces and actuates the neural tissue to design an efficient, durable, and above all, a safe interface.

2.3.1 Mechanism of Charge Transfer

Charge is carried by electrons in stimulation/electrical circuits and electrodes. As observed before, ions like Na^+, K^+, Cl^-, etc., are the charge carriers in the physiological medium/neural tissue. So when an electrode is placed inside a physiological medium/neural tissue, an interface is formed between the two mediums and a process of transduction of charge carriers from electrons (in electrode) to ions (in tissue) occurs. The process of charge transfer although complicated (it depends on stimulation paradigms, material properties, tissue characteristics, etc.) can be classified into two mechanisms, one in which no electron transfer occurs at the interface and another in which electrons are transferred at the interface. The former is called non-faradaic reaction wherein a capacitive mechanism is at work, and the latter is called faradaic reaction wherein reduction or oxidation reactions at the interface are responsible for electron transfer.

If the stimulation requires only a small amount of charge to be delivered, the charge redistribution occurs such that there is no transfer of electrons across the interface. When the stimulation polarity is reversed, the charge redistribution is also reversed. Such an interface with non-faradaic processes can be modeled as a simple capacitor.

If the stimulation requires charge transfer that is greater than that supported by non-faradaic processes, electrons may be transferred by oxidation and reduction reactions at electrodes which are driven positive and negative, respectively, during the stimulation cycle. The byproducts of these reactions cannot be recovered when the direction of current is reversed if the byproducts are allowed the time to diffuse away from the interface. In an interface model, the faradaic processes are represented by a variable resistor [9]. Figure 2.8 shows the equivalent circuit of the electrode–tissue interface.

Typically, the reactions are either mass transport limited or reaction kinetics limited. With the former, due to fast kinetics and slow transport of material, the reactants do not move away from the interface and there is an effective storage of charge. As a result, some or all reactants are available for a reverse reaction when the polarity of the stimulation pulse is reverse. The degree of availability determines the degree of reversibility of the reactions. With reaction kinetics limited reactions, the reactant do move away from the interface, are dissolved into the tissue and are unavailable for reverse reactions when the polarity of the stimulation pulse is reversed. The dissolution of products into the tissue changes its environment and has a damaging effect.

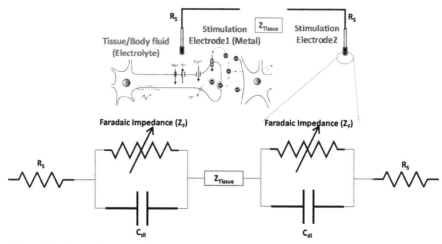

Figure 2.8   Two electrode–tissue interfaces are formed during bipolar stimulation between two electrodes with the tissue in between. Each component of the interface (Z_F, C_{dl}, and Z_{Tissue}) can have non-linear properties that may be dependent on the stimulation conditions. Capacitive processes are denoted by C_{dl} and the faradaic processes (reversible and irreversible) are represented by Z_F. R_S represents the impedance of the wire connecting the stimulation source like an implantable neurostimulator and the electrode. R_S, Z_F, and C_{dl} can be modified by appropriate selection of electrode material and surface topography.

During stimulation, as the electrode is driven away from its equilibrium potential, all charge flows through the capacitive mechanism. As the potential increases, the charge begins to be transferred by the Faradaic mechanism of oxidation or reduction. An extreme case is near the so-called water-window. In the physiological medium, water is available in abundance and is not mass limited. At very high and low potentials, water undergoes oxidation and reduction, respectively, resulting in formation of oxygen and hydrogen gas in the medium through irreversible faradaic reactions. These potentials form the upper and lower limits of any interface. The potential cannot exceed beyond these limits and the range defined by them is called the water window. The goal of any interface design is to avoid any irreversible faradaic reactions during the electrical stimulation paradigms of interest.

2.3.1.1 Charge injection

Constant current charge injection and constant voltage charge injection are the two most common methods of charge injection for neurostimulation applications. In constant current mode, a constant current source is attached between cathode and anode (with tissue in between them). This is the most common method in neurostimulation. In constant voltage mode, a voltage source is applied between the cathode and anode. Irrespective of the mode, there are three categories of charge injection: monophasic, biphasic with balanced charge, biphasic with imbalanced charge.

Charge recovery is better with biphasic stimulation and hence, it is preferred. The type of electrode material used and the type of electrolyte/physiological environment, would dictate the type and level of irreversible faradaic reactions at the interface due to positive and negative potential during a biphasic pulse. These reactions will dictate delivery of charge balanced or imbalanced biphasic pulse.

2.3.2 Electrode Properties and Materials

The material must be biocompatible and should not evoke significant foreign body response (any material elicits some response consisting of encapsulation by macrophages, microglia, fibroblasts, astrocytes, endothelial, and meningeal cells [10]), the material must have application-appropriate mechanical properties (ability to penetrate the meninges for micro-electrode applications, ability to withstand multiple years of mechanical bending for spinal-cord applications), it should be able to withstand mechanical stress imposed during the surgical procedure, the electrode material should be able to sustain adequate charge delivery to elicit the required neural response without undergoing toxic (irreversible) faradaic reactions or corrosion, and function-sustaining conducting and insulating properties of the material should be maintained during the life of the implant. The above requirements are significantly higher for chronically implantable systems used in clinical applications as compared to acute systems used for animal research studies.

Electrodes used for neural stimulation can be divided into two categories based on their surface area: macro-electrodes (area $> 100,000$ μm^2) and micro-electrodes (area $< 10,000$ μm^2) [11]. Macro-electrodes are less susceptible to corrosion-induced degradation because they have lower charge-densities/phase and higher charge-delivery/phase. The latter has implications on tissue viability and is covered in the next section. Micro-electrodes are more susceptible to corrosion induced degradation due to higher charge-densities/phase, but have a lower charge-delivery/phase. Use of macro- or micro-electrodes is application-specific. Majority of the commercial clinical applications such as deep-brain stimulation, cochlear implants, and spinal cord stimulation utilize macro-electrodes.

Over the last few decades, numerous materials have been evaluated for their ability to inject charge by capacitive and faradaic mechanisms and their overall suitability (biocompatibility, manufacturability, etc.) for neural stimulation. Iridium Oxide, Platinum (Pt) and Platinum Iridium (PtIr) alloy, tantalum, and titanium nitride are examples of materials that have been reported in the literature for use in electrode materials. To increase the effective surface area of the electrodes, deposition techniques like sputtering is employed to produce porous geometries [12]. Extensive research is ongoing in the development and evaluation of electrically conducting polymers such as Polyethylenedioxythiophene (PEDOTs) and carbon nanotubes as alternatives to metallic elements used in electrodes. These new materials provide the capability of chemical surface modification with physiologically active species to minimize tissue response, and potentially improve chronic functionality of the electrodes.

2.3.3 Limits to Electrical Stimulation

Studies have shown that electrical stimulation to the tissue can also induce damage to it. Although the precise manner in which stimulation induced tissue damage occurs is not fully understood yet, two classes of mechanisms have been put forth. In one line of theory and as indicated earlier, it is possible to produce reaction byproducts due to irreversible faradaic processes at the electrode-tissue interface during cathodic stimulation. If the rate of these reactions is greater than what can be tolerated by the tissue, damage can occur. In another line of thought, neuronal hyperactivity or their firing for an extended period of time changes the local environment. This causes depletion in intracellular and extracellular concentration of oxygen, glucose, and other ions resulting in cell death. Excessive release of excitatory neurotransmitters like glutamate may also cause excitotoxicity. For electrodes interfacing with the peripheral nerves, mechanical tightening/compression around the nerves may also cause damage [13].

Numerous studies have been performed to study the impact of electrical stimulation on the tissue. In [14], Shannon presented an expression for the maximum safe level of stimulation as

$$\log\left(\frac{Q}{A}\right) = k - \log(Q).$$

Where Q = charge per phase (μC per phase), Q/A = charge density per phase (μC/cm^2 per phase), and k is a constant. No tissue damage is observed

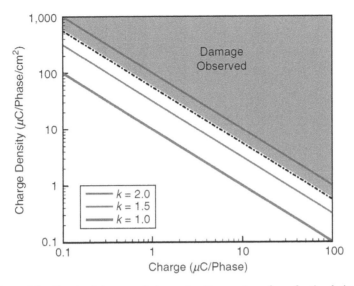

Figure 2.9 Graph of charge and charge density per phase for safe stimulation.

Source: https://commons.wikimedia.org/wiki/File:Shannon_Plot.png

if $k < {\sim}1.5$ and it is observed if $k > {\sim}1.7$. The Figure 2.9 graphically illustrates this.

Finally, standard stimulation protocols like the use of biphasic stimulation wherein is charge delivered to the tissue during one phase is recovered subsequently is essential to reduce damage to the tissue.

It is generally best of keep the pulse-width narrow to reduce any electro-chemical reactions on the electrode surface. However, the narrowness may be dictated by the amount of current that can be delivered by the stimulator. Specifically, as the duration of the pulse (pulse-width) reduces, the corresponding constant current pulse amplitude required for activation increases. This relationship is called the strength–duration curve and is shown in Figure 2.10.

The equation

$$I_{th} = \frac{I_{rh}}{1 - \exp(-\frac{W}{\tau_m})}.$$

Where I_{th} = threshold current at each pulse-width, I_{rh} = rheobase current, W = pulse-width, τ_m = membrane time constant, describes the strength–duration relationship. The definitions of rheobase and chronaxie are explained in the caption of the Figure 2.10.

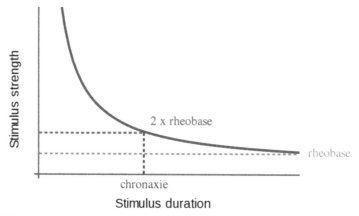

Figure 2.10   Strength–duration curve shows the required stimulation intensity at a given stimulus duration to elicit action potential. Rheobase is the current intensity when using an infinitely long pulse-width. Chronaxie is the pulse-width duration required when using a current intensity that is twice the rheobase.

Source: https://commons.wikimedia.org/wiki/File:Rheobase_chronaxie.png

2.4 Discussion: Bringing it All Together

Over the years, researchers have successfully utilized new stimulation paradigms and electrode configurations to elicit novel stimulation effects for research and clinical purposes. A few examples are presented next to give the readers a glimpse into what is possible in terms of neurostimulation if electrodes and stimulation paradigms are applied in a manner that utilizes principles of electricity and physiology.

To achieve stimulation at a distance from the electrode site, the concept of virtual electrodes is employed. Typically, if an electrode is driven positive (anodic stimulation) the neuronal membrane under it will be hyperpolarized. This hyperpolarization and movement of ions may in turn cause depolarization at locations away from the electrode and thus initiate action potentials. Typically, the anodic stimulation amplitude required to elicit an action potential is three or more times higher than cathodic stimulation amplitude.

To achieve an increase in stimulation threshold, the concept of accommodation is employed. Here, a subthreshold cathodic stimulation pulse is delivered resulting in subthreshold depolarization. Such subthreshold depolarization increases sodium inactivation thereby reducing the number of channels that are available for activation which increases the stimulation threshold.

Conversely, to achieve a momentary reduction in stimulation threshold, a long-duration (compared to sodium channel dynamics) hyperpolarizing pulse is delivered that desensitizes the partial inactivation mechanism of the sodium channel. Upon removing the hyperpolarizing pulse with an abrupt rectangular pulse, for example, the sodium activation gate conductance increases quickly relative to the inactivation gate resulting in a period of time when there is a net in-flow of sodium ions accompanied by membrane depolarization which may result in an action potential. This approach is called anodic break. Slowly rising pulses like a triangular or an exponential may not be as effective in achieving anodic break as a rectangular pulse.

Achieving selectivity during neurostimulation may be its biggest challenge. Selectivity implies stimulating a certain set of neurons without stimulating others in the same spatial location. With conventional neurostimulation, large-fiber diameter neurons with greater distance between Nodes of Ranvier are stimulated first. This is in contrast to the physiological case wherein smaller fiber neurons are activated first. To address this challenge, Fang and Mortimer [15] utilized a hyperpolarizing pulse which has greater influence on the large diameter fibers compared to small diameter fibers to desensitize the former. The subsequent stimulation pulse caused selective and earlier activation of small diameter fibers compared to the large diameter fibers. Protocols with selective activation of interneurons and nerve terminals have also been reported [16].

Neuromodulation may be one of the most exciting interdisciplinary fields of research and clinical application in this decade. It presents the opportunity to interface, innervate and thus influence the functioning of one of the greatest mysteries in science and one of the greatest challenges in medicine; the human brain. This chapter provided an overview of the physiological, electrical and material principals essential to achieve neuromodulation. This basic understanding may help readers with expertise in engineering, and life sciences to think of efficient and efficacious modulation circuits and systems including interfaces that fully exploit and better augment the capabilities of an intact or compromised nervous system for improved research and clinical outcomes.

References

[1] Wichmann, T., and DeLong, M. R. (2006). Deep brain stimulation for neurologic and neuropsychiatric disorders. *Neuron* 52, 197–204.

[2] McIntyre, C. C., and Grill, W. M. (1999). Excitation of central nervous system neurons by nonuniform electric fields. *Biophys. J.* 76, 878–888.

[3] Goldman, D. E. (1943). Potential, impedance, and rectification in membranes. *J. Gen. Physiol.* 27, 37–60.

[4] Hodgkin, A. L., and Katz, B. (1949). The effect of sodium ions on the electrical activity of the giant axon of the squid. *J. Physiol.* 108, 37–77.

[5] Hodgkin, A. L., and Huxley, A. F. (1952). A quantitative description of membrane current and its application to conduction and excitation in nerve. *J. Physiol.* 117, 500–544.

[6] Paralikar, K. J., Rao, C. R., and Clement, R. S. (2009). New approaches to eliminating common-noise artifacts in recordings from intracortical microelectrode arrays: inter-electrode correlation and virtual referencing. *J. Neurosci. Methods* 181, 27–35.

[7] van Dongen, M., and Serdijn, W. (2016). *Design of Efficient and Safe Neural Stimulators*. Berlin: Springer.

[8] McIntyre, C. C., Miocinovic, S., and Butson, C. R. (2007). Computational analysis of deep brain stimulation. Expert review of medical devices, 4(5), 615–622.

[9] Merrill, D. R., Bikson, M., and Jefferys, J. G. R. (2005). Electrical stimulation of excitable tissue: design of efficacious and safe protocols. *J. Neurosci. Methods* 141, 171–198.

[10] Polikov, V. S., Tresco, P. A., and Reichert, W. M. (2005). Response of brain tissue to chronically implanted neural electrodes. *J. Neurosci. Methods* 148, 1–18.

[11] Cogan, S. F. (2008). Neural stimulation and recording electrodes. *Annu. Rev. Biomed. Eng.* 10, 275–309.

[12] Weiland, J. D., Anderson, D. J., and Humayun, M. S. (2002). *In vitro* electrical properties for iridium oxide versus titanium nitride stimulating electrodes. *IEEE Trans. Biomed. Eng.* 49, 1574–1579.

[13] McCreery, D. B., et al. (1992). Damage in peripheral nerve from continuous electrical stimulation: comparison of two stimulus waveforms. *Med. Biol. Eng. Comput.* 30, 109–114.

[14] Shannon, R. V. (1992). A model of safe levels for electrical stimulation. *IEEE Trans. Biomed. Eng.* 39, 424–426.

[15] Fang, Z, and Mortimer, J. T. (1991). Selective activation of small motor axons by quasitrapezoidal current pulses. *IEEE Trans. Biomed. Eng.* 38, 168–74.

[16] McIntyre, C. C., and Grill, W. M. (2002). Extracellular stimulation of central neurons: influence of stimulus waveform and frequency on neuronal output. *J. Neurophysiol.* 88, 1592–604.

3

Front-End Sensing Amplifiers

You Zou

Verily Life Science, US

Abstract

Neural sensing amplifiers are essential blocks for acquiring bio-signals in neuroscience and clinical research for treating a variety of diseases related to neural disorders. Such amplifiers are used to extract and condition micro- to millivolt scale signals aroused from weak neural activities. As such, they typically have ultra-low noise level as they usually dominate the noise level of front-end sensing blocks. Further, miniaturized neural amplifiers based on integrated circuits offer benefits of small form factors and ultra-low power consumption, allowing unobtrusive operations and minimum tissue heat dissipation. They are, hence, gaining popularity rapidly for wearable and implant applications. This chapter discusses example designs and circuit considerations for low-power and low-noise neural sensing amplifiers suitable for portable and implantable medical devices.

3.1 System Level Architecture

The architecture discussed in the frequently-cited bio-sensing amplifier developed by Harrison and Charles [1] has been widely adopted. The schematic of the proposed neural amplifier is shown in Figure 3.1.

This architecture contains an operational transconductance amplifier (OTA) with capacitive voltage feedback scheme to circumvent the DC offset voltage caused by electrochemical effects at the electrode–tissue interface. The close-loop gain is set by

$$A_{\text{close loop}} = \frac{C_{\text{in}}}{C_{\text{fb}}}. \tag{3.1}$$

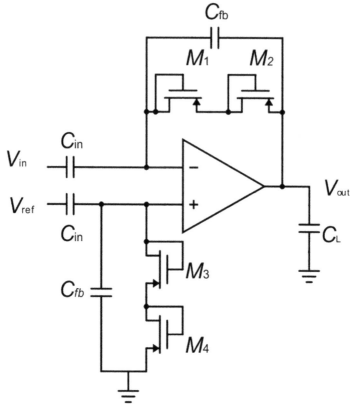

Figure 3.1 Capacitive-coupled neural amplifier architecture [1].

Pseudoresistors realized by metal-oxide-semiconductor (MOS) transistors are utilized to provide large resistance in a relatively small area. For voltage across the pseudotransistor less than 0.2 V, resistance greater than 100 GΩ has been reported.

Assuming the parasitic capacitance at the input of the OTA is C_p, the input-referred noise of this bio-amplifier can be written as a function of the input-referred noise of the OTA and the feedback network as follows.

$$\overline{V_\mathrm{ni,amp}^2} = \left(\frac{C_\mathrm{in} + C_\mathrm{fb} + C_\mathrm{p}}{C_\mathrm{in}}\right)^2 \times \overline{V_\mathrm{ni,OTA}^2}. \tag{3.2}$$

Typically C_in is much greater than C_fb. As a result, the input-referred noise of the neural amplifier is dominated by that of the OTA.

3.2 Current Mirror-Based OTA

Harrison and Charles [1] have discussed many practical circuit design techniques for bio-signal sensing OTAs. The OTA described in this work is a conventional current mirror-based OTA with wide output swing shown as Figure 3.2.

Assuming all transistors are in strong inversion and operating in saturation region. The input-referred thermal noise of this OTA can be found as

$$\overline{V^2_{\text{ni,thermal}}} = \left[\frac{16kT}{3g_{\text{m1}}} \left(1 + 2\frac{g_{\text{m3}}}{g_{\text{m1}}} + \frac{g_{\text{m7}}}{g_{\text{m1}}} \right) \right] \times \triangle f. \qquad (3.3)$$

In Equation (3.3), $\triangle f$ is the noise bandwidth and it is equal to $\frac{\pi}{2} \times \mathbf{BW}$, where $\mathbf{BW}$ is the –3 dB bandwidth of the OTA. In order to minimize the input-referred noise, one needs to make sure that $g_{\text{m1,2}}$ are maximized while $g_{\text{m3,7}}$ are minimized. This can be achieved by operating the input devices in deep weak inversion region while pushing the diode-connected load into strong inversion. It, however, should be noted that the stability of the amplifier is the limiting factor of how far $g_{\text{m3,7}}$ can be minimized. This is because the poles associated with the capacitance at the gate of $g_{\text{m3,7}}$ equal to $\frac{g_{\text{m3,7}}}{C_{\text{g}}}$, which will

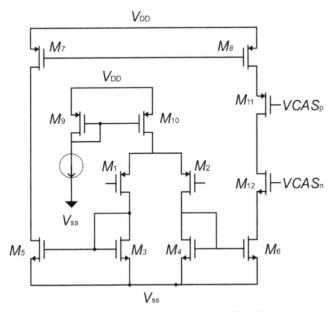

Figure 3.2 Current mirror-based OTA [2].

become so close to the dominant pole to $\frac{g_{m1,2}}{C_L}$, so that the phase margin may become insufficient.

As for minimizing the flicker noise, the input transistors should be made as large as possible. However, C_3 and C_7 increases with the device size, compromising the phase margin of the amplifier as analyzed in the previous paragraph. Moreover, the increase of input device sizes will increase the input-referred noise according to Equation (3.2). As a result, there exist optimum input device sizes which minimize $\frac{1}{f}$ noise.

For low-noise OTAs, the noise efficiency factor (NEF) is often used to quantify the trade-off between noise and power consumption. The NEF is introduced in Steyaert and Sansen [2] and written as

$$\text{NEF} = V_{ni,rms}\sqrt{\frac{2I_{tot}}{\pi \times U_T \times 4kT \times \text{BW}}}. \tag{3.4}$$

Where $V_{ni,rms}$ is the input-referred rms noise voltage, I_{tot} is the total current consumption, U_T is the thermal voltage, BW is the amplifier bandwidth in Hertz. Substituting Equation (3.3) into Equation (3.4) and assuming $g_{m3,7} \ll g_{m1}$, we have

$$\text{NEF} = \sqrt{\frac{16}{3U_T}\left(\frac{I_{D1}}{g_{m1}}\right)}. \tag{3.5}$$

Where I_{D1} is the drain current through m1. NEF can hence be minimized by minimizing $\frac{I_{D1}}{g_{m1}}$, which can be realized by biasing the input devices in weak inversion, where $g_{m1} = \frac{\kappa I_{D1}}{U_T}$. Also, since the input-referred thermal noise of each input device biased in weak inversion region can be estimated as $\frac{2kT}{\kappa g_m}$. Assuming a typical value for $\kappa = 0.7$, Equation (3.4) thus becomes

$$\text{NEF} = \sqrt{\frac{4}{\kappa^2}} \cong 2.9. \tag{3.6}$$

It has thus been found that the theoretical limit of NEF is approximately 2.9 for this topology.

3.3 Folded-Cascode OTA

The neural amplifier designed by Wattanapanitch et al. [3] used a modified folded-cascode topology aiming at increasing the power efficiency. The topology is shown in Figure 3.3.

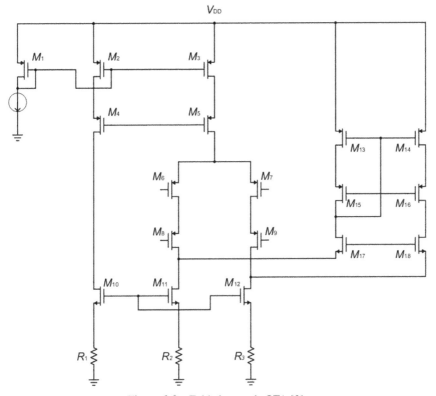

Figure 3.3 Folded-cascode OTA [3].

The high-power efficiency is achieved by scaling down the current sup-plied to folded branches aggressively. The noise contribution by the devices at folded branches can be mitigated at the same time. The current sources are cascoded to ensure accurate current scaling.

There is, however, an issue introduced by such severe scaling. In con-ventional folded-cascode amplifiers without such current scaling, the current going through the folded branch is close to that of the input devices since the input impedance to the folded branch is relatively low compared with the current source biasing the input devices. Therefore, in this case, the overall effective transconductance G_m will be close to that of the input device. In contrast, current scaling will result in much less current going through the folded branch. As a result, the overall transconductance G_m is much smaller than that of the input devices due to a much higher impedance looking into the cascade branch. This will degrade the noise efficiency since G_m is significantly less for the same total current consumption.

In order to overcome this, source degeneration resistors are used in the input branch current sources. The output impedances of the cascoded input-differential pair and the source-degenerated current sources are made much larger than the impedance looking into the sources of the M_{17} and M_{18}, which will therefore make G_m of the entire OTA to near g_{m1}. In addition, the noise contributions from the source-degenerated current sources are dominated by the resistors which can be made much smaller than the noise by MOS current source without source generation.

Assuming the input devices are the only noise sources of the OTA, the input-referred noise of the ideal OTA can be approximated as

$$V_{\text{ni,rms}} = \sqrt{\frac{4kT \times U_{\text{T}}}{\kappa^2 \times I_{\text{D}}} \times \frac{\pi}{2} \times \text{BW}}. \tag{3.7}$$

Where I_{D} is the DC current going through each source degeneration resistor. Therefore, the theoretical limit of the NEF for such a topology can be found as

$$\text{NEF} = \sqrt{\frac{2}{\kappa^2}} \cong 2.02. \tag{3.8}$$

Compared with the work by Harrison and Charles [1], the NEF of this topology is improved due to the saving of power consumption in the folded branch while still maintain the same level of input referred noise. Nevertheless, the large source-degeneration resistor (on the order of $M\Omega$) will increase the die area and also voltage overhead.

Another neural recording amplifier designed by Qian et al. [4] uses current splitting technique to enhance G_m of the folded-cascode OTA topology. The topology proposed is shown in Figure 3.4.

Assume the bias current in each half circuit is I_B, which is divided into $N = A + B$ fractions.

In this topology, the small signal current of M_1 adds in-phase with that of M_3, and assume this current flows into M_5 due to the current sink source degeneration technique discussed. Similarly, the small signal current of other half-circuit due to M_2 and M_4 flows into M_6. As a result, assuming sub-threshold operation, the effective overall transconductance can be written as

$$G_m = \frac{\kappa I_B}{U_{\text{T}}} \times \frac{2A + 1}{N}. \tag{3.9}$$

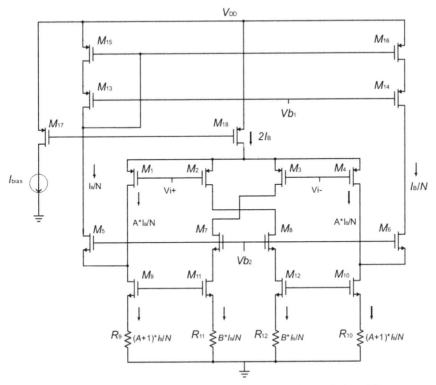

Figure 3.4   Folded-cascode OTA with current-splitting technique [4].

It can be observed that G_m is proportional to A. However, a large G_m will not result in lower input-referred noise because the output current noise power spectral density (PSD) due to M_2 and M_3 will also be scaled up by $\left(\frac{A+1}{B}\right)^2$. In fact, the input-referred noise voltage PSD of this topology can be found in Equation (3.10).

$$\overline{V_{ni}^2} = \frac{4kT}{U_T} \times \left[A + B\left(\frac{A+1}{B}\right)^2\right] \times \frac{I_B}{N} \times \frac{1}{G_m^2}. \qquad (3.10)$$

As a result, the optimal for A to minimize input-referred noise level can be found equal to $N/2$. The theoretical NEF limit can also be found comparable to that without the current splitting.

3.4 Complementary-Input OTA

Complementary-input OTAs are gaining its popularity rapidly for neural recording applications mainly due to its better noise-power performance compared with conventional architectures.

Holleman et al. [5] introduced an ultra-low power biopotential amplifier using such a topology. The amplifier schematic is shown in Figure 3.5.

The input devices are both biased to operate in sub-threshold region to maximize the $g_m/$ID ratio. The input devices are AC-coupled which has provided the flexibility for biasing the PMOS and NMOS devices separately using pseudoresistors.

Similarly, assuming the complementary input devices are biased in sub-threshold region, the theoretical NEF for this topology can be found as

$$NEF = \frac{1}{\sqrt{2\kappa}}\ldots \tag{3.11}$$

It can be observed that the NEF of this topology is $\frac{1}{2\sqrt{2}}$ times of that of the topology by Harrison [2]. There are mainly two design techniques for this topology that can be attributed to the reported excellent noise-power efficiency. First, this design uses only one current branches (compared with four current branches in the design by Harrison [2]). This ensures minimal power consumption. Also, it minimizes the number of devices that would

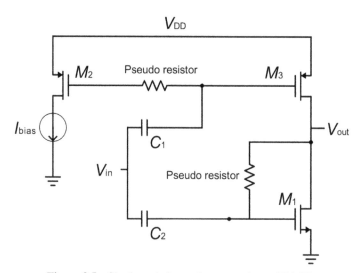

Figure 3.5   Single-ended complementary input OTA [5].

contribute to input-referred noises as those in the folded stage of folded-cascode amplifiers. Second, inverter-based input devices are adopted. In this case both PMOS and NMOS devices are driven by input signals so that the g_m/ID ratio has effectively been doubled. As a result, the input-referred noise power is reduced by half.

This design has achieved excellent noise-power efficiency. On the other hand, it also has several flaws. First, the linearity is poor by using open-loop architecture. The authors have to use digitally controlled diode-connected MOS at the output in order to provide relatively well-controlled linearity. Second, the power rejection ratio (PSRR) also requires much improvement in a system where a clean power line is not available. In this design, the source of the input devices is directly connected to the supply. Their gates are also coupled to the power supply through gate-source parasitic capacitors. As a result, the PSRR is only approximately 6 dB. This poor PSRR will, therefore, make output signals susceptible to the noise from power supply, which will corrupt the signal in the band of interest.

The PSRR, linearity and accuracy of this topology can be significantly improved by utilizing a closed-loop fully differential architecture as described by Zhang et al. [6]. The OTA topology depicted is shown in Figure 3.6.

The input stage of this topology is basically the fully differential version of the previous design. It also utilized complementary input devices, which doubles the effective G_m and hence, reduces the input-referred noise voltage by a factor of $\sqrt{2}$ compared with traditional OTAs. The OTA is used in

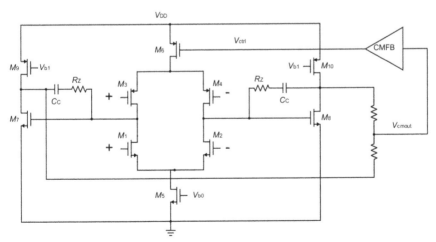

Figure 3.6 Fully-differential OTA with complementary input stage.

a feedback network formed by the capacitive divider which increases the input-referred noise at the input of the OTA by C_{fb}/C_{in}, which is typically 10%. Assuming $g_{m1} = g_{m3}$, the input-referred thermal noise power can be estimated as

$$\overline{V_{ni}^2} = \left(\frac{16kT}{3g_{m1} \times 2} \right) \times \Delta f. \tag{3.12}$$

Similar, this topology reduces the input-referred noise voltage by $\sqrt{2}$ compared with traditional fully differential OTAs due to the G_m doubling. The largest advantage of this topology compared with the single-ended approach in Figure 3.5 is that in this design dual tail current sources are used in the first stage to ensure high CMRR and PSRR.

The common-mode gain can be expressed as

$$A_{cm} \simeq \frac{(g_{o5} + g_{o6})g_{m8}}{g_{o1}g_{o2}[1 + \frac{sC_c}{g_{o5}+g_{o6}}]} \tag{3.13}$$

Note that g_{o1} and g_{o2} are output conductance of the first and the second stage, and g_{m8} is the transconductance of transistor M_8. C_c is the compensation capacitor. g_{o5} and g_{o6} denote the transconductance of transistor M_5 and M_6, respectively.

The power supply interference can be found as

$$A_{ps} = \frac{V_{out}}{V_{in,supply}} \simeq \frac{\gamma \Delta g_m \frac{g_{m8}}{g_{m1}g_{m2}}}{1 + \frac{sC_c}{\Delta g_m}}. \tag{3.14}$$

Where γ is the attenuation of the supply variation before being amplified by the transconductance mismatch in M_3 and M_4, and it can be written as

$$\gamma = \frac{g_{m6}}{g_{m3} + g_{m4}} \times \left(1 - \frac{V_{g6}}{V_{dd}} \right) \tag{3.15}$$

The CMFB contains a wide band and small gain amplifier since the gain from the output of this small gain amplifier to V_{cmout} is sufficiently large. Let $g_{m,i}$ denotes the transconductance of corresponding transistor M_i, C_c, and C_L denote the compensation and load capacitors, g_{o1} and g_{o2} denote the total output conductances of the first and the second stage. We have the expressions for the common mode feedback path gain and different mode gain as

$$A_{cmfb} = \frac{V_{out,CM}}{V_{ctrl}} = \frac{-sg_{m6}C_c + g_{m6}g_{m7,8}}{s^2 C_c C_L + sC_c g_{m7,8} + g_{o1}g_{o2}} \tag{3.16}$$

$$A_{cmfb} = \frac{V_{out,DM}}{V_{in,dm}} = \frac{-s(g_{m1,2}+g_{m3,4})C_c + (g_{m1,2}+g_{m3,4})g_{m7,8}}{s^2 C_c C_L + sC_c g_{m7,8} + g_{o1}g_{o2}}. \tag{3.17}$$

Clearly, the same compensation capacitor needs to be tuned to ensure the stability of both common-mode and differential-mode paths, similar to conventional fully differential amplifiers with common-mode feedback.

There are a number of topology variations based on the complementary-input architecture in order to further decrease the power consumption. Song et al. [7] proposed an OTA that is a hybrid of complementary-input and folded cascode topology. The schematic is shown in Figure 3.7.

Pseudoresistors are again used to provide bias voltages. This circuit drives both the NMOS and PMOS input pair to exploit the benefits of doubling effective G_m. Further, the supply voltage of the input branch and the folded branch is provided separately, which allows aggressive scaling down of the supply voltage to the input branch and, therefore, saving power. The bias voltages for the NMOS and PMOS input devices are also provided separately which reduces the minimum input stage supply voltage from $2V_{GS} + V_{DS,min}$ to $3V_{DS,min}$. The minimum input stage supply voltage is reported as low as 0.3 V. Compared with the complementary-input telescopic topology shown in Figure 3.6 whose minimum power supply is $2V_{GS} + 2V_{DS,min}$,

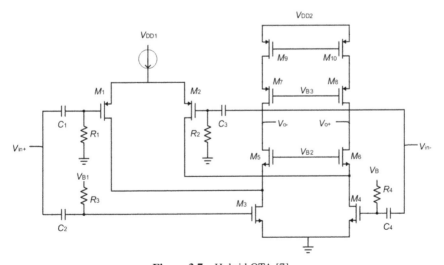

Figure 3.7 Hybrid OTA [7].

the save on power consumption due to the supply voltage scaling-down is significant.

3.5 Current Reuse OTA

Song et al. [7] proposed a stacked current-reuse folded cascode topology. The schematic is shown in Figure 3.8.

This topology contains two independent folded cascode amplifiers sharing the same mid-rail current sink/source (MCS) which is highlighted in a dashed box in Figure 3.8. The MCS is a key block in this design. The schematic of this block is shown in Figure 3.9.

The MCS contains two PMOS and NMOS current mirrors with their source connected. The gate voltages of the current mirrors are dictated by the input bias current and the reference input voltage to the negative feedback amplifier on the left. The voltage across the MCS needs to be large enough (greater than $2*V_{DS,min}$) to ensure it is in saturation region.

Compared with a conventional folded cascode amplifier, this topology again exploits the benefits of driving the complementary in out devices and

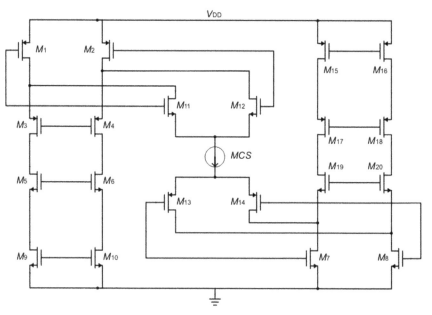

Figure 3.8   Current-reuse OTA [7].

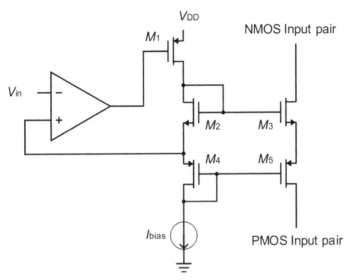

Figure 3.9   Mid-rail current source [7].

hence the effective gm is doubled. Further, the current consumption per channel is reduced by half since the same bias current is reused by two channels. In tins case, the NEF of this topology is reduced by half compared with that of a single differential pair.

On the other hand, the minimum supply voltage for this topology is $4V_{GS} + 2V_{DS,min}$, which is $2V_{GS}$ higher than that of the complementary-input amplifier as shown in Figure 3.6 which has been discussed by Zhang et al. Another drawback of this architecture is that the noise level for the two amplifiers cannot be tuned separately, which is required by some applications such as fetal ECG monitoring as indicated by Song et al. [7].

3.6 Reconfigurable OTA

Programmability is a crucial feature for versatile neural sensing amplifiers that can be used to sense different types of bio-signals. Such an amplifier typically provides tunable bandwidth and gain to accommodate different requirements during recording different neural activities such as EEG, ECG, and EMG, etc. In addition, programmability is also desired in order to overcome the process variation and mismatch issues.

For capacitive-coupled neural amplifiers, the midband gain is given by $\frac{C_{in}}{C_{fb}}$ as discussed earlier. As such, the midband gain is often controlled by digitally tuning the input and/or feedback capacitance. For example, in the work done by Lopez et al. [8], three-bit digitally-selectable parallel input capacitors are used as shown in the Figure 3.10.

As another example, the biopotential sensing amplifier designed by Wang et al. [9] chooses to digitally switch in and out the feedback capacitance to adjust the gain.

Configurable bandwidth of neural sensing amplifiers can also be achieved in multiple ways. For the widely used capacitive-coupled neural amplifier architecture shown in Figure 3.1, it typically has two poles to have a band-pass characteristic. The lower-frequency pole is determined by the feedback resistors and capacitors as follows.

$$p_1 = \frac{1}{C_{fb} R_{fb}}. \tag{3.18}$$

Where C_{fb} is the feedback capacitor and R_{fb} is the feedback resistor. The higher-frequency pole can be estimated as

$$p_2 = \frac{G_m}{C_{Load} A_{CL}} \tag{3.19}$$

Where G_m is the effective transconductance of the OTA, A_{CL} is the close-loop gain and C_{Load} is the load capacitance.

Based on Equation (3.18), tuning the feedback resistor and capacitor can both tune the low-frequency pole. However, switching in and out capacitors in the feedback capacitor array will change the mid-band gain as well. As a result, tuning the resistance value of the feedback pseudoresistor is typically employed. This is usually realized by controlling the gate voltage

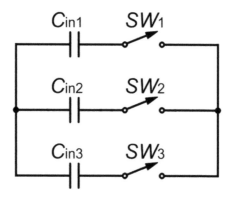

Figure 3.10   Tunable capacitor array for reconfigurable midband gain [8].

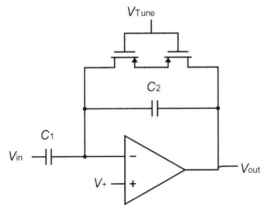

Figure 3.11 Tunable feedback resistance for lower-frequency pole tuning [10].

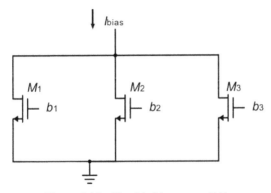

Figure 3.12 Tunable bias current[11].

of the transistors that form the pseudoresistor. For example, Sodagar et al. [10] implemented the band-tunable neural amplifier as shown below. The schematic of this implementation is shown in Figure 3.11. Programmable bandwidth from 100 Hz to 10 KHz has been reported. The resistance is however a nonlinear function of the control voltage.

In order to tune the higher-frequency pole, the load capacitance and/or the transconductance can be adjusted. The former has been demonstrated by Lopez et al [10]. The load capacitors are implemented as programmable capacitor arrays to facilitate bandwidth programming. This approach, however, does not help with power consumption. On the other hand, the OTA bias current can be adjusted to tune the transconductance and hence vary the bandwidth of the neural amplifier. This approach allows DC power saving.

It has been implemented by Zou et al. [11]. The schematic of the tunable bias current sources is shown in Figure 3.12. The bias current sources are controlled by three-bit digital signals.

References

[1] Harrison, R. R., and Charles, C. (2003). A low-power low-noise CMOS amplifier for neural recording applications. *IEEE J. Solid State Circuits* 38, 958–965.

[2] Steyaert, M. S. J., and Sansen, W. M. C. (1987). A micropower low-noise monolithic instrumentation amplifier for medical purposes. *IEEE J. Solid State Circuits* 22, 1163–1168.

[3] Wattanapanitch, W., Fee, M., and Sarpeshkar, R. (2007). An energy-efficient micropower neural recording amplifier. *IEEE Trans. Biomed. Circuits Syst.* 1, 136–147.

[4] Qian, C., Parramon, J., and Sanchez-Sinencio, E. (2011). A micro-power low-noise neural recording front-end circuit for epileptic seizure detection. *IEEE J. Solid State Circuits* 46, 1392–1405.

[5] Holleman, J., and Otis, B. (2007). "A sub-microwatt low-noise amplifier for neural recording," in *Proceedings of the 29th Annual International Conference of the IEEE Engineering in Medicine and Biology Society*, Lyon, 3930–3933.

[6] Zhang, F., Holleman, J., and Otis, B. P. (2012). Design of ultra-low power biopotential amplifiers for biosignal acquisition applications. *IEEE Trans. Biomed. Circuits Syst.* 6, 344–355.

[7] Song, S., et al. (2015). A low-voltage chopper-stabilized amplifier for fetal ECG monitoring with a 1.41 power efficiency factor. *IEEE Trans. Biomed. Circuits Syst.* 9, 237–247.

[8] Mora, C., Lopez et al. (2012). A multichannel integrated circuit for electrical recording of neural activity, with independent channel programmability. *IEEE Trans. Biomed. Circuits Syst.* 6, 101–110.

[9] Wang, T. Y., Lai, M. R., Twigg, C. M., and Peng, S. Y. (2014). A fully reconfigurable low-noise biopotential sensing amplifier with 1.96 noise efficiency factor. *IEEE Trans. Biomed. Circuits Syst.* 8, 411–422.

[10] Sodagar, A. M., Perlin, G. E., Yao, Y., Najafi, K., and Wise, K. D. (2009). An implantable 64-channel wireless microsystem for single-unit neural recording. *IEEE J. Solid State Circuits* 44, 2591–2604.

[11] Zou, X., Xu, X., Yao, L., and Lian, Y. (2009). A 1-V 450-nW fully integrated programmable biomedical sensor interface chip. *IEEE J. Solid State Circuits* 44, 1067–1077.

4

Circuits for Neural Stimulation

Peng Cong

Verily Life Science, US

4.1 Introduction

In typical neuromodulation, therapy is achieved by delivering electrical energy from an implantable device into neural system. This energy is typically in the form of pulse train as illustrated in Figure 4.1.

The stimulation pulse can be in constant voltage mode or constant current mode. Therapy delivery, or stimulation pulse delivery, from a circuit point of view, is one of the key functions within the neuromodulation device. It transfers energy from a charge storage element, e.g., a battery or capacitor, to the tissue. The interface between the stimulating electrode(s) in an implantable device and the target tissue forms a complex electrochemical interface as discussed in Chapter 2. To the first order, the interface can be modeled as a capacitor and a resistor in series with a resistor in parallel with the capacitor, as illustrated in Figure 4.2. For simplicity, the resistor can be omitted for analysis. Although it is a complicated system, over the years, researchers have elucidated various electrode designs and stimulation patterns that avoid or limit unintended tissue damage and electrode corrosion during charge delivery [1].

The design objective is to ensure that the charge delivery circuit is capable of interacting with all elements of the stimulation system, including the target tissue and lead, to delivery stimulation waveform with the targeted waveform shape, amplitude, frequency, and pulse width while meeting goals of safety, efficiency, and efficacy.

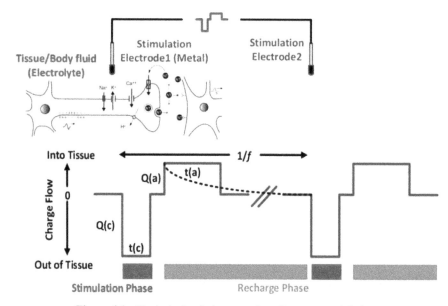

Figure 4.1   Typical stimulation waveform for neuromodulation.

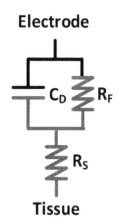

Figure 4.2   Simplified electrode–tissue interface.

4.2 Stimulation Waveform

Most classical stimulators use constant current stimulation instead of constant voltage stimulation. As stimulation intensity is typically closely related to the charge injected into tissue, the stimulation current determines the intensity of the stimulation for a given pulse width. Stimulation intensity is directly controllable by current-based stimulation. Another related reason is that using this approach the amount of charge injected can be relatively easily controlled.

As shown in Figure 4.1, anodic and cathodic charge delivery are in half-phases, for which the total charge delivered during the anodic phase reverses most – if not all – of the charge delivered during the cathodic phase. The anodic phase may consist of passive or active recharge phases, or a combination. Passive recharge is enabled by shorting two electrodes together. Typically, passive recharge is involved a coupling capacitor, which is often placed in series between the implantable device output and the electrode. One function of this capacitor is to avoid a potential path for DC current flow into the tissue. During passive recharge, charge is allowed to flow into the tissue in a decaying exponential fashion set by the product of the electrode impedance and the combination of the coupling capacitor, C_{DC}, and the double-layer capacitor, C_D, in the electrode model. During active recharge, charge is actively injected across the electrode–tissue interface for the purpose of rapidly reversing potentials on the electrode so as to limit tissue damage, or allow for balanced-charge high rate stimulation [2]. Figure 4.3 illustrates a simplified schematic of the stimulation circuit interfacing with tissue. C_{DC} is a DC block capacitor to ensure no DC current into tissue and C_{DC} is typically designed to be much larger than C_D, the double-layer capacitor in the electrode interface model.

From a power efficiency point view, passive recharge (dash line in Figure 4.1) is more efficient than active recharge (solid line in Figure 4.1), because no current from battery supply is needed for passive recharge. However, as stimulation frequency increases, passive recharge, relying on the RC constant of the electrode interface, may not be sufficient for charge balance without causing much bias over the stimulation capacitor, where active recharge is more effective.

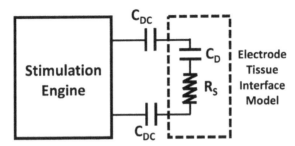

Figure 4.3 Simplified schematic of circuit for stimulation therapy.

Besides constant current stimulation, it is also possible to use constant voltage stimulation. In this case, the current through the tissue is not only a function of stimulation voltage but also tissue interface as well, which sometime is not well controlled and can vary over time. However, voltage stimulation could be more efficient as there is no voltage drop on constant current source. Therefore, most pacemakers use a constant voltage instead of a constant current, where power consumption is more critical.

Besides classical stimulation schemes using a constant current or constant voltage, several studies, however, have shown that other waveforms can lead to more effective stimulation. The stimulation waveform can benefit from the dynamic properties of the membrane as described by the Hodgkin–Huxley equations [3, 4]. It was shown that Gaussian stimulation waveforms need less energy to obtain the same neural recruitment compared to rectangular waveforms [3]. Furthermore, the waveform also has influence on the charge injection capacity of the electrodes. In [4] an algorithm is presented to find the optimal stimulation waveform from an energy efficiency point of view and again Gaussian waveforms are found [5].

Another characteristic of the stimulation waveform is bipolar and unipolar stimulation. As illustrated in Figure 4.4, two active current sources/sinks are required for bipolar stimulation and only one active current source/sink is required for unipolar stimulation. For bipolar stimulation, typically two electrodes are involved. One electrode and a device can are used for unipolar stimulation, where the impedance of the can can be much smaller than that of an electrode.

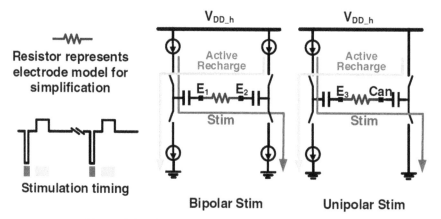

Figure 4.4   Bipolar and unipolar stimulation waveform.

4.3 Circuit Design for Stimulation

4.3.1 Circuit Design

Figure 4.5 presents an simplified schematic of the stimulation circuit, where C_{DC} and C_D are removed as the voltage drop on those components is typically low for safety reasons. Typically, the impedance of the electrode interface is 1 kΩ (Z_{load}). The stimulation current could be as large as a few tens of mA for some applications. The supply voltage for stimulation circuit, V_{DD_h}, would be well beyond 10 V, which is much more than a typical battery voltage in 1–5 V range. Therefore, a voltage boosting circuit is required.

The high voltage can be designed to be sufficient for worst case stimulation parameters; however, as illustrated in Figure 4.6, the design could be very insufficient when stimulation current amplitude or tissue impedance changes. When the product of stimulation current and electrode impedance is high, the efficiency could be high and only limited by the overhead voltage to ensure adequate operation of current source. However, when the product is small due to stimulation amplitude decreasing or electrode impedance decreasing, the efficiency will drop significantly. Therefore, a dynamic adjustment of high voltage supply is required to improve this system-level efficiency as illustrated in Figure 4.7. Ideally, V_{DD_h} tracks the required voltage with an overhead voltage to ensure current source in adequate operation region.

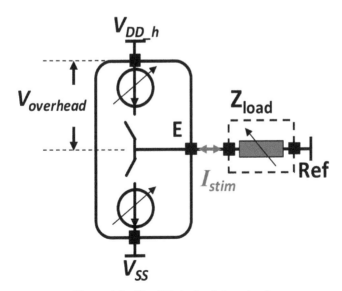

Figure 4.5 Simplified stimulation circuit.

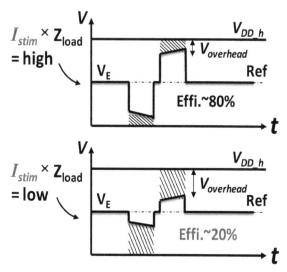

Figure 4.6   Illustration of power efficiency drop with stimulation amplitude and electrode impedance decrease.

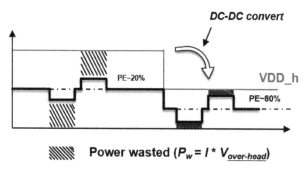

Figure 4.7   Power efficiency improvement with dynamic high voltage adjustment.

A charge pump with tunable scaling is one solution to generate high voltage supply for the stimulation circuit. The idea of charge pump is to charge an individual capacitor to battery voltage (sometimes to charge two series-connected capacitors to battery voltage for 0.5x of battery) and then stack them up for high voltage supply. A typical range for neuromodulation device is from 1.5x to 5x of battery voltage, which ranges typically from 3 to 4 V. A separate capacitor with large value is also typically used as a holding capacitor to provide relative stable supply. Typically, no regulator is used for power efficiency consideration. The charge pump configuration with scaling

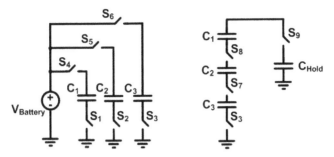

Figure 4.8 Charge pump design example with gain of 3x.

of three is presented in Figure 4.8 as an example. The operation needs two phases, charge phase and pump phase, to complete. During the charge phase, capacitors C_1, C_2, and C_3 are connected in parallel to the battery through switches. Each capacitor gets charged up to the battery voltage. During the pump phase, capacitors C_1, C_2, and C_3 are connected in series with each other and in parallel to the hold capacitor and the hold capacitor gets pumped up to 3x the battery voltage. Sometime the capacitor can be stacked on battery to achieve an extra 1x voltage gain. Other voltage scaling values can be achieved with variations in the capacitor and switching configurations. Due to the energy and efficiency requirements, the capacitors in the charge pump are external capacitors with the μF range and the charge pump stack is dynamically adjusted to bias the supplies to the minimal operating point as discussed previously.

The next important element in the stimulation circuit is a current source or current sink, as shown in Figure 4.5, driving the issue interface. There are two major considerations that necessitate a precision current source or sink. The first is the therapeutic requirement. The stimulation amplitude could range from 100 μA to a few tens of mA. It is critical to have a precision current source and sink to achieve this requirement. The second is for charge balance, where the charge injection during stimulation phase should match the charge recovered in the recharging phase. At the same time, the overhead voltage needs to be minimized for high efficiency. There are many techniques for precision current generation. Fundamentally, they are typically involved with current mirror design. One of the design examples is shown in Figure 4.9. In this design example, the reference current described above is sampled, then amplified and passed to the tissue using an active mirror system. Since the current source and sink circuits are complementary, only the description of the sink is presented below. In this design, a feedback sense resistor-based

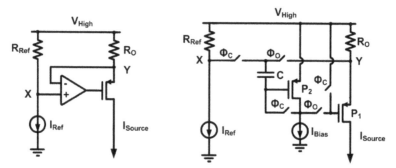

Figure 4.9   Example of stimulation circuit with improved current matching.

architecture is utilized for two primary reasons: (1) The architecture allows the current matching to be driven by resistors rather than transistors. This results in better matching in most integrated circuit processes and, (2) The architecture eliminates the need to keep any output transistors in saturation, reducing voltage headroom requirements to improve efficiency under many conditions of neurostimulation. As the left figure in Figure 4.9, to achieve accurate mirroring, the nodes X and Y are forced to be the same by the negative feedback of the amplifier, which results in the same voltage drop on R_{Ref} and R_O. Therefore, the ratio between output current, I_{Source}, and the reference current, I_{Ref}, equals the ratio of R_{Ref} and R_O.

In a typical process, the current matching is limited to roughly eight bits by residual mirror errors due to amplifier offsets unless excessive voltage is placed across the resistors, which would undermine efficiency. The circuit shown in Figure 4.9 addresses this issue using a correlated double sampling (CDS) switched-capacitor circuit with a single transistor as an amplifier. The circuit operation involves two phases: the precharge phase and the output phase. In the precharge phase, Φ_C is high and Φ_O is low. P_2 is diode-connected and C stores the voltage difference between Node X and the gate voltage of P_2. P_1 is off in this phase. In the output phase, Φ_C is low and Φ_O is high. The capacitor, C, retains the voltage acquired in the precharge phase. When the voltage at Node Y is exactly equal to the earlier voltage at Node X, the stored voltage on C biases the gate of P_2 properly so that it balances I_{Bias}. If, for example, the voltage across R_O is lower than the original R_{Ref} voltage, the gate of P_2 is pulled up, allowing I_{Bias} to pull down on the gate on P_1, resulting in more current to R_O. In this design, charge injection is minimized by using a large holding capacitor of 10 pF. The performance is eventually limited by resistor matching, leakage, and finite amplifier gain.

The stimulation rate can be tuned from 0.153 Hz to 1 kHz and the pulse-width can be tuned from 100 µs to 12 ms, which covers the key operating space of most neuromodulation research. However, the actual limitation in the stimulation output pulse-train characteristic is ultimately set by the energy transfer of the charge pump and safety limitation of electrodes.

The ratio of R_{Ref} to R_O is typically in the order of 100. However, for large output dynamic, it may not be enough and require reference current (I_{Ref}) to be adjustable. One of the implementation of current source is based on R–2R. As shown in Figure 4.10, a stable current is used as reference for the system and a constant current of 100 nA, generated and trimmed on chip, is used to drive the reference current generator, which consists of an R–2R-based DAC to generate an 8-bit reference current with a maximum value of 5 µA. An on-chip sense-resistor-based architecture was chosen for the current output stage to eliminate the need to keep output transistors in saturation, reducing voltage headroom requirements to improve power efficiency. The architecture uses thin-film resistors (TFRs) in the output driver mirroring to enhance matching.

The proposed design can also be used for other applications, e.g., driving LED or µ-LED for optogenetic. One of the implementation is shown in Figure 4.11, where multiple channels are combined together to provide a large driving current for optogenetics [10].

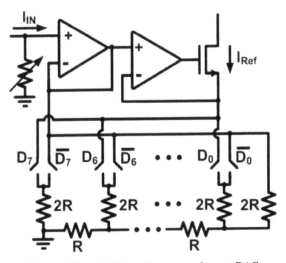

Figure 4.10 R–2R-based current reference DAC.

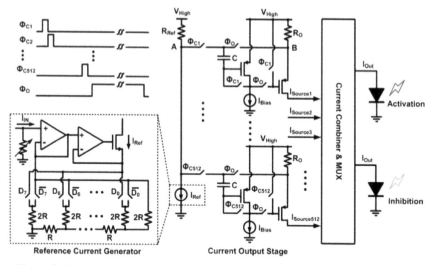

Figure 4.11   An implementation using stimulation circuit for optogenetic application.

4.3.2 Safety Consideration

4.3.2.1 Charge balance

As mentioned previously, it is important to keep charge balance stimulation for safety reasons. For example, in Figure 4.5, the charge delivered in t(c) is required to be the same as the charge recovered during t(a). Although the current reference and current mirror can be precise, extra care needs to be taken to ensure charge balance. No matter how precisely the stimulation current matches recharge current, there is always a mismatch of currents and the mismatch could potentially accumulate. The stimulation circuit design can not only rely on circuit matching.

Over the years, there are many novel designs to address this challenge. As presented in Figure 4.12, to ensure charge balance and safety, the potential at each active electrode is compared with the reference potential. The output of the control unit decides if the electrode voltage exceeds a safe window. If it is positive, a short current pulse in micro seconds, e.g., 5 μs, will be applied to the electrodes. As a consequence, the electrode potential is steered toward a more balanced condition. The process will continue until the electrode has achieved a predefined voltage window around the reference voltage. This method is controlled, guarantees safe operation, and gives feedback information on the electrode condition after a stimulation. The process can be applied to all the electrodes until all charge balancers verify that all electrodes

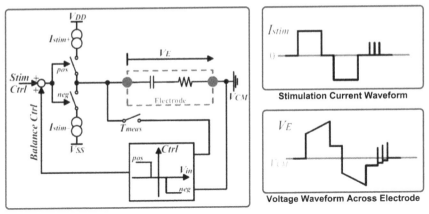

Figure 4.12 Example of an active charge balancing design [6].

are safe, and this finishes this stimulation cycle [6]. One of the challenges for this approach is the unknown biological consequence with the small stimulation pulses and the process may take some time to finish.

Another example is illustrated in Figure 4.13. The idea is to combine with active recharge and passive recharge. Active recharge phase ensures charge balance as good as possible by using a dynamic current-balancing circuit to match a pMOS current source with an nMOS current sink. Then any residual charge will be handled by passive recharge by shorting electrodes together.

A biphasic current pulse is initiated by a control signal. Then a sample-and-hold (S/H) phase begins. In this phase, the pMOS gate voltage required to support an nMOS current is sampled on a hold capacitor as shown in Figure 4.13(a). I_{bs} and M_2 form a source–follower that diode-connects transistor M_1, and is biased with enough current to drive at a reasonable speed. The source–follower also level-shifts V_{PO} to a diode drop below V_P, which keeps the active-cascode formed by M_3 and M_4 in saturation. After the circuit has settled, SAMP1 is opened first. SAMP is then opened followed by the closing of the HOLD switch as shown in Figure 4.13(a). This circuit in Figure 4.13(b) incorporates a closed-loop switched-capacitor sample-and-hold [8], which attenuates noise and other disturbances to V_P by the gain of the loop around transconductor of S/H. Errors due to the charge injection of the SAMP1 switch but not the SAMP switch limit the precision of the sample and hold [8]. Any residual charge will be handled by a passive recharge phase by closing the switch in parallel with tissue impedance as shown in Figure 4.13(b).

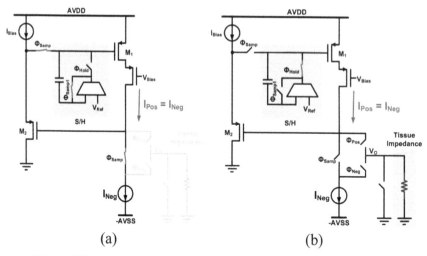

(a)   (b)

Figure 4.13   Combination of active and passive charge balancing design [7].

4.3.2.2 Electrode safety consideration

Safety is the most important consideration for nerve stimulation. Even with charge balance, the neural interface needs future study. The stimulation waveform is presented in Figure 4.14. Also assume charge balanced stimulation, which means that the total charge injected into tissue during the stimulation phase is totally recovered during the recharge phase [Q(c) = Q(a)].

When an electrode is inserted into tissue, an electrochemical interface would be formed as a simplified model shown in Figure 4.15. During stimulation, the double-layer capacitor (C_D) is charged up, as the V_D shown in the simplified model.

When V_D is small, no reactive happens. It is just a pure capacitor and the charge will be recovered totally during the recharging phase. When V_D is getting larger, the "real" current (Faradaic current) goes through the double

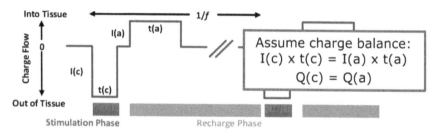

Figure 4.14   Stimulation waveform in constant current mode with charge balance.

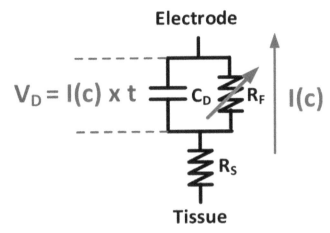

Figure 4.15   Voltage over double-layer capacitor during stimulation.

layer, electrochemical reaction happens, and there could be some electrode material dissolved into tissue fluid through electrochemical reaction. Electrochemical reaction primarily depends on the electrode material and V_D. Therefore, for a given electrode material, V_D should be minimized to avoid those reactions. One way of minimizing V_D is to use small stimulation current and small pulse width. However, stimulation current and stimulation pulse width are largely determined by therapeutic requirements. The other way is to increase the capacitance of the double-layer capacitor of the electrode by the electrode's geometry and to modify surface condition. Therefore, electrode design is extremely important for efficient and safe stimulation.

Whether it is safe or not largely depends on whether the reaction is reversible or irreversible. Figure 4.16 presents three different cases. In the figure, the top is an illustration of the electrochemical reaction process and the bottom is the responding potential waveforms (the top one is current waveform and the both is the voltage/potential across C_D, the double layer capacitor). Figure 4.16(a) only involves a double-layer capacitor acting as a pure capacitor, when V_D is small. Electrons and ions come to the electrode–tissue interface but nothing across. During this process, the voltage potential (V_D) across C_D increases linearly with constant current stimulation.

The second column, Figure 4.16(b), involves a double-layer capacitor and reversible reaction. In this case, platinum (Pt) is used for illustration. During the process, some electrons will go across the interface to generate really current (Faradaic current). Electron, hydrogen ions, and platinum will react

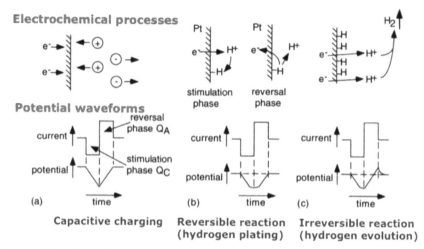

Figure 4.16   Electrochemical processes and potential waveforms during charge-balanced stimulation: (a) capacitive charging only; (b) reversible hydrogen plating; (c) irreversible hydrogen evolution.

with a product of Pt and hydrogen bond. However, the unique property of this product is that they only stay at the electrode surface. This process is typically called hydrogen plating. The potential waveform across C_D will be a little different from the pure capacitor case. At the beginning, it acts as a double capacitor, but when V_D reaches a level where hydrogen plating happens, the slope of the V_D is much flat. From an electrical modeling perspective, although electron transfer occurs, in terms of the electrical model, it still can be modeled as a capacitor with much bigger value, $\sim$10x for Pt. Therefore, it is called a pseudo-capacity capacitor. As the products did not go away from the electrode, e.g., Pt, all the product can be recovered during the recharging phase if it is a charge balanced stimulation. It is generally considered safe and this is why Pt is widely used for electrodes (note, typically Pt alloy is used for mechanical consideration).

The last column involves a double-layer capacitor and irreversible reaction, which need to be avoided. In this case, some electrons will go across the interface to generate really current (Faradaic current) and the products diffuse away from the electrode. If the products diffuse away from the electrode, the charge cannot be recovered upon reversing the direction of current. One example is water reduction, where hydrogen will be formed and it is irreversible. V_D for water reduction is about 1 V.

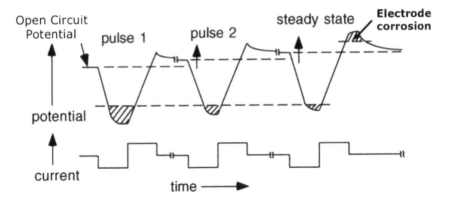

Figure 4.17 Issues with charge balance stimulation for irreversible stimulation.

Typically, water reaction should be avoided and it would get worse if stimulation still keeps charge balance. As illustrated in Figures 4.16(c) and 4.17, the potential across C_D will not come back to 0 and the accumulated charge will bring it to more positive, which introduces electrode corrosion. At steady state the net charge will introduce constant electrode corrosion. In this case, charge balance is not desired per safety consideration, although charge-balanced stimulation is critical to ensure safety for a reversible case.

Another way to do it is to monitor the electrode potential instead of monitoring the charge. The stimulator can be designed to control the voltage cross double-layer capacitor within a limitation. For the reversible case, it also ensures charge balance.

4.3.2.3 Neuron safety consideration

Beside the electrode safety considerations, neuron safety is an even bigger topic and people still do not have a complete understanding to determine neuron safety under stimulation. Shannon criteria are widely accepted standards as illustrated in Figure 1.8, proposed by professor Shannon in 1992. Note the criteria are purely experimental by plotting previous results against charge per phase (x-axis) and charge densities (the charge divided by the area of the electrode) per phase (y-axis). The line in Figure 1.8 is defined by the equation below:

$$k = \log(Q) + \log(Q/A), \tag{4.1}$$

where Q is charge per phase and A is the area of electrode. If K is greater than 1.5–1.8, it would not be safe. Shannon criteria have been evolved in recent years for microelectrodes.

4.4 Summary

Stimulation is one of the most important and critical elements in implantable neuromodulation devices. The following is a summary of circuit considerations for stimulation:

- Work with electrode designer and understand electrode model
- Small DC leakage current
- Charge balanced stimulation for reversible reaction
- Monitor and control voltage across double-layer capacitor within "good" reaction window if irreversible reaction might happen.
- Limit maximum charge and charge density for neuron safety
- Constant current vs. constant voltage stimulation
- Power-efficient design (typically, the major power consumption source in implantable devices).

References

[1] Durand, D. B. (2006). "Electrical stimulation of excitable systems," in *Biomedical Engineering Fundamentals*, ed. J. D. Bronzino (Boca Raton, FL: CRC Press).

[2] Yoo, H.-J., and van Hoof, C. (2011). *Bio-Medical CMOS ICs*. Boston, MA: Springer.

[3] Sahin, M., and Tie, Y. (2007). Non-rectangular waveforms for neural stimulation with practical electrodes. *J. Neural Eng.* 4:227.

[4] Wongsarnpigoon, A., and Grill, W. (2010). Energy-efficient waveform shapes for neural stimulation revlealed with a genetic algorithm. *J. Neural Eng.* 7:046009.

[5] van Dongen, M., and Serdijn, W. (2016). *Design of Efficient and Safe Neural Stimulators*. Boston, MA: Springer.

[6] Ortmanns, M., Rocke, A., Gehrke, M., and Tiedtke, H.-J. (2007). A 232-channel epiretinal stimulator ASIC. *IEEE J. Solid State Circ.* 42, 2946–2959.

[7] Sit, J.-J., and Sarpeshkar, R. (2007). A low-power blocking-capacitor-free charge-balanced electrode-stimulator chip with less than 6 nA DC error for 1-ma full-scale stimulation. *IEEE Trans. Biomed. Circ. Syst.* 1:3.

[8] Martin, K., and Johns, D. A. (1997). *Analog Integrated Circuit Design*. Toronto, ON: Wiley.

[9] Merrill, D. R., Bikson, M., and Jefferys, J. G. R. (2005). Electrical stimulation of excitable tissue: design of efficacious and safe protocols. *J. Neurosci. Methods* 141, 171–198.

[10] Paralikar, K., Cong, P., Yizhar, O., Fenno, L. E., Santa, W., Nielsen, C., et al. (2011). An implantable optical stimulation delivery system for actuating an excitable biosubstrate. *IEEE J. Solid State Circ.* 46, 321–332.

5

Embedded Signal Analysis

Z. Wang and N. Verma

Princeton University, US

Abstract

Medical devices are advancing to provide increasingly intelligent and relevant responses, bringing the potential for greater impact and better patient outcomes. Such responses are derived from analysis of patient signals made available thanks to advances in sensing and instrumentation technologies. However, given the complexity of the underlying physiologic and pathophysiologic processes, analysis of these signals is challenging on many levels. Data-driven methods for analysis are showing substantial promise towards these challenges, actuated both by the large-scale emergence of data in the medical domain and by progress in algorithms from the machine-learning and statistical-signal-processing domains. This chapter explores the platform-level challenges with extending data-driven methods to resource-constrained (energy, area) wearable and implantable devices. Recent work in the area of heterogeneous microprocessor design and low-power algorithms is presented to illustrate approaches to addressing the challenges.

5.1 Introduction

Recent advances in instrumentation technologies and sensing technologies (many of which are covered in this book) have given us the ability to acquire a range of patient signals, particularly in medically relevant contexts (patient behaviors, time scales, etc.). However, utilizing such signals toward advanced control functions in medical devices also requires embedded capabilities for signal analysis. Generally, patient-signal analysis faces several distinct challenges, which are overviewed in this chapter. Thus, before proceeding, it is worth identifying the key tradeoffs between localized signal analysis within

71

a medical device, where resource constraints such as energy are generally substantial, and remote signal analysis away from a medical device, where resource constraints are generally more relaxed. Typically, these tradeoffs arise due to communication energy, bandwidth, and latency for transmitting data to remote platforms, especially using wireless technologies. While generally the tradeoffs must be analyzed for each application, we find that many cases emerge where localized signal analysis is preferred.

As an example, in this chapter we will pay special attention to energy, which is found to be a primarily constraint in implantable and wearable medical devices, due to battery size [1]. Table 5.1 compares the power breakdown of two seizure-detection systems based on sensing and analyzing electroencephalogram (EEG) signals [2]. The first system sends sensed EEG data wirelessly to a centralized processor for all the digital signal analysis, while the second system performs local feature extraction, using a specialized digital processor, before transmitting the data. With the high energy cost of wireless communication, employing such a computation vs. communication tradeoff to reduce communication enables substantial reduction of overall power consumption ($15\times$) Computation vs. communication tradeoffs can have a similar benefit in systems limited by communication bandwidth (e.g., inductively coupled transcutaneous data transmission [3]) and latency (e.g., closed-loop therapeutic stimulation [1]).

However, to exploit local computation in this way, it must be as energy efficient as possible. For instance, while local classification in the system of Table 5.1 would have further reduced communication, the algorithm employed, if simply targeted to an embedded microprocessor, would have increased the local power consumption by 1–2 orders of magnitude [2]. The focus of this chapter is to explore the many opportunities available for enhancing the energy efficiency of local signal analysis, from the algorithms level to the circuits level. In particular, this will be done in the context of

Table 5.1 Power comparison for EEG-based seizure-detection systems with and without local processing [2]

	Wireless EEG	Local Feature-Extraction
I-Amp array	72 μW	72 μW
ADCs (12b, 11kS/s)	3 μW	3 μW
Digital processors	–	2.1 μW
Radio (CC2550) [11]	1733 μW	43 μW
• Active: *bit-rale* × 40nJ/bit	• Active: 43.2kbps × 40nJ/bit	• Active: 2kbps/2sec × 40nJ/bit
• Start-up: 4.8 μW	• Start-up: 4.8 μW	• Start-up: 4.8 μW/2sec
• Idle mode: 0.46 μW	• Idle mode: 0.46 μW	• Idle mode: 0.46 μW/2sec
Total	1808 μW	120 μW

the application-level challenges arising in the case of specific patient signals, which are showing increasing value in medical devices.

5.2 Challenges in Medical-Signal Analysis

In computational medicine, increasingly, rich indicators of disease, disease states, and disease progression are being discovered in patient signals derived from complex underlying physiologic and pathophysiologic processes. Generally, such signals are challenging to analyze. Advances in data-driven methods of modeling and analysis, combined with the availability of data in the medical domain, have opened up unprecedented capabilities for addressing some of the challenges. Before delving into the algorithmic methods and computational platforms for enabling these in wearable and implantable devices, we overview some of the high-level challenges typically encountered:

1. **Complex processes:** The complexity of underlying physiologic and pathophysiologic processes, and the complex mechanisms by which these processes are expressed in the signals we can sense, makes analysis difficult. This is illustrated in the EEG signal of an epileptic patient shown in Figure 5.1. As an example, we see that simple anomaly

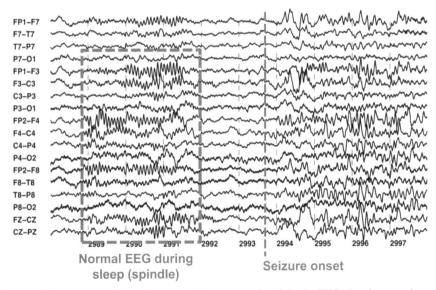

Figure 5.1 Distinguishing seizure onset from normal activity in EEG signals necessitates high-order models.

detection for identifying seizures would be inadequate. Normal physiologic *spindle* activity (which commonly occurs during sleep) in fact can have substantial resemblance to seizure activity, necessitating higher-order rules for signal inference. While substantial progress has been made in modeling cellular and cellular-circuit dynamics leading to seizures, such models typically are not adequate for discriminating between the range of signal activities expressed in the signals, particularly when observed through equally complex transduction processes leading to the signals.

2. **Patient-to-patient variability:** Beyond the underlying complexity of the patient signals, necessitating high-order models for analysis, matters are further complicated by the variability from patient to patient [4, 5]. Such variability is prominent both in the expression of physiologic processes and in the expression of targeted pathophysiologic processes, both of which impact the ability to perform effective discrimination. Thus, there is often the need for patient-specific models for signal analysis, and thus the need for scalable approaches to construct such models. As an example, Figure 5.2 shows EEG signals corresponding to seizure onset from two patients. We see significant variation, both spatially in the particular EEG channels distributed around the scalp which exhibit a response, and spectrally in the frequencies which serve as biomarkers of the response.

3. **Dynamic Processes:** Adding to the complexities identified above is the challenge that the underlying processes are dynamic. This is especially true following acute events. As an example, Figure 5.3 shows the electrocardiogram of a patient following a myocardial infraction [6, 7]. During this time the heart undergoes substantial changes, which

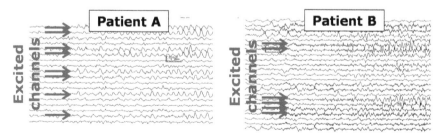

Figure 5.2 Substantial patient-to-patient variability is seen in EEG expressions of physiologic/pathophysiologic processes.

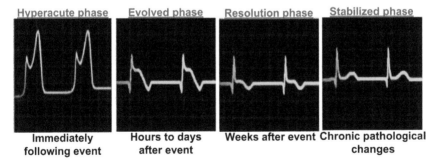

Figure 5.3 Changes expressed in ECG following a myocardial infraction [6].

impact its operation and thus that signals arising from its operation. As seen, such changes can be large in comparison to the relatively minute variances corresponding to the signal-analysis models employed for discriminating targeted activity. Thus, the models must evolve with such dynamics.

5.3 Data-Driven Medical-Signal Analysis

Taking into consideration the challenges discussed above, this section turns attention to data-driven approaches for creating models for signal analysis. In particular, while analytical/mechanistic approaches based on modeling the underlying physics of the processes have shown success in various instances [8], for many systems such models alone will not be feasible. On the other hand data-driven methods are showing unprecedented success in a broad range of medical applications [9].

This has been actuated by two factors. First, medical practices of today have resulted in large-scale availability of patient data. This is through the various electronic bed-side monitors employed within hospitals, the ubiquitous use of electronic records, etc. Second, advances in data analysis, from the domains of machine learning and statistical signal processing have given powerful tools for utilizing that data toward a range of medical inferences. Indeed, the convergence of these two factors had led to broad-spanning transformations in healthcare, for improving understanding, diagnosis, and treatment of diseases. Thus, aside from the success that data-driven methods have shown, motivation for employing these within wearable and implantable devices also comes from the potential to integrate more broadly with modern and future healthcare methods.

To illustrate the promise of data-driven methods for addressing the challenges mentioned above, consider the example of detecting cardiac arrhythmias based on ECG signals, shown in Figure 5.4. First, with regards to the complexity of underlying processes, the precise biomarkers such a signal expresses may be latent (or certainly not obvious) to clinical experts. Data-driven methods have provided tools for identifying signal features that exhibit correlation to processes of interest, enabling the discovery of signal biomarkers. Following from this, the relation between signal biomarkers and processes of interest may also be unknown, particularly in a statistical sense taking into account various other processes which are occurring simultaneously, causing superimposing activity in the signals. Data-driven methods provide tools for modeling the relationship between identified biomarkers (e.g., some mapping of the signal features, as shown) and such processes, in the presence of practical variances occurring in the signals (e.g., through the decision boundary, as shown). Second, with regards to patient-to-patient variability, selective use of data, perhaps from an individual or from a patient group defined by some similarity metrics, data-driven methods provide systematic and efficient tools for constructing customized models for analysis. Third, with regards to dynamic processes, the envisioned ability to acquire data on an ongoing basis (e.g., through emerging technologies for chronic monitoring, the potential arises for data-driven models to track changes expressed in the signals. We point out that while protocols for doing this have not yet been established, technologies and algorithms based on data-driven methods show significant potential in this regard.

Thus, data-driven methods are promising on several levels for analyzing medical signals. However, their application in resource-constrained scenarios, such as within wearable and implantable devices, raises distinct technical challenges. The remainder of this chapter explores the structure and

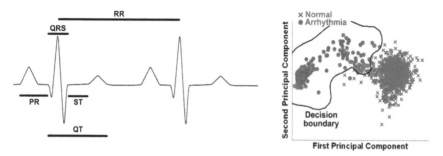

Figure 5.4   Data-driven methods for ECG-based cardiac arrhythmia detection.

attributes of signal-analysis algorithms based on data-driven methods, and then goes on to look at circuits and architectures that exploit this towards resource-efficient implementations.

5.4 Overview of Inference Systems

This section overviews the general structure of a signal-inference system. As shown in Figure 5.5, such a system consists of two stages: (i) feature extraction; and (ii) inference. In the case of data-driven inference, the operations performed by either or both of these stages is based on a model formed from previous observations of data, referred to as a *training dataset* (e.g., figure shows inference model derived from training via a training dataset). The subsections below address each of these stages.

5.4.1 Feature Extraction

The objective of feature extraction is to transform input signals into a representation that facilitates pattern recognition, corresponding to an inference of interest. Often, this implies two practical considerations for the transform. The first consideration is to create a representation whereby similarity metrics between instances of data can be defined for use by the inference stage. For instance, a common approach to feature extraction is to transform signals to a Euclidean vector space, possibly of high dimensionality. Many machine-learning algorithms then employ vector distances to represent how similar two instances of data are. The second consideration is to create a representation that enhances generalization of learning, by resulting in maximal separation between instances of data, with respect to the inference of interest. Essentially, this corresponds to suppressing variances in the signal that

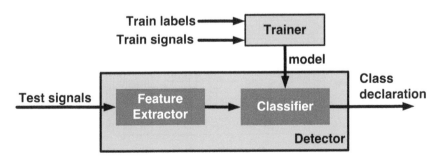

Figure 5.5   Generic structure of a data-driven inference system.

have low correspondence with the inference, and accentuating variances that have high correspondence with the inference. For instance, there are many cases where the process of interest is not characterized by a particular time reference; in such cases, features that are time-shift invariant are typically chosen to suppress time dependence. Together, these two considerations enable instances of data to be evaluated in terms of how similar they are to previously observed instances of data and how, with high confidence, previously observed instances of data map to specific inferences.

For illustration, we examine spectral-energy distribution, which is a generic biomarker of the EEG and, therefore, a widely utilized feature set for signal analysis [10]. Figure 5.6 shows a signal-processing system, which may be employed for feature extraction. Here, each EEG channel is fed to a bank of band-pass filters, whose outputs are accumulated (over a predefined epoch length) after taking the absolute value, to extract the energy. The energy from each filter, over all of the channels, then forms a feature vector.

The suitability of spectral-energy distribution as a feature set for signal inference can be seen in Figure 5.7, where EEG data from an epileptic patient is employed to detect the onset of seizures. Two different feature sets are shown. The first one corresponds to broadband features, primarily aiming to achieve time-shift invariance, while the second corresponds to spectral-energy features, according to Figure 5.6 (though the feature vectors over 18 EEG channels are of high dimensionality, they have been projected to two dimensions via principal-component analysis simply for visualization).

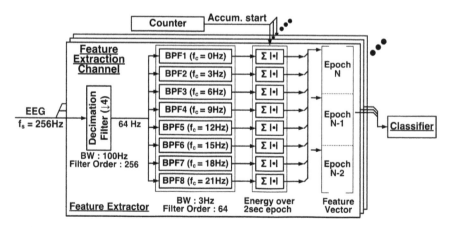

Figure 5.6   Seizure detection system with EEG spectral energy distribution as features.

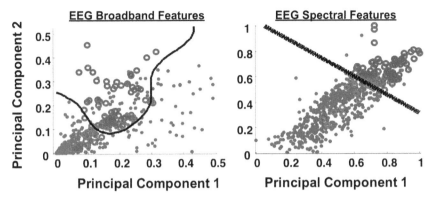

Figure 5.7 Comparison of two different feature sets for EEG-based seizure detection.

With seizure and non-seizure data shown by different markers, we see that the spectral-energy features achieve much better separation between the distribution of the two classes. The benefit this has is that a comparatively simpler and more robust decision boundary can be formed, as a model for making inferences in the subsequent stage. The following section will raise the concern of generalization error, which arises with data-driven models; specifically, an inference model can fit to the training dataset, but if the training dataset does not adequately represent the statistics of future data, the inference performance will be poor. Complex distributions, as resulting from the broadband feature set are more susceptible to generalization error, necessitating larger training datasets to adequately represent the statistics. This can be problematic in medical applications, where such data may be not be available, especially from a particular patient (for customization) and especially corresponding to infrequent pathophysiologic processes.

Incidentally, generalization error points to another key objective of feature extraction, which is dimensionality reduction. While the raw data may correspond to a large number of time samples, and thus high dimensionality, generally the size of training dataset required to adequately reduce the potential for generalization error increases with feature-vector dimensionality [11]. Thus, it is typically preferable to employ as small a number of features as possible, and if the number of informative features is high, often feature construction [12] and/or feature selection [13] will be applied.

Having overviewed some objectives for feature extraction, the question that follows is how suitable features are determined for a particular application. Very commonly in medical-sensor applications, the choice of features is driven by domain knowledge, for instance an understanding of likely biomarkers within a signal. However, taking the perspective of designing platforms for medical signal analysis, a critical consideration is that the precise choice of features depends on a number of factors. Certainly, feature sets are variable across applications and application signals, but indeed features are also variable among patients, for a particular application. As an example, keeping with seizure detection, in additional to EEG spectral-energy distribution, features associated with heart-rate variability have shown to improve detection performance in some patients [14].

An alternative to features determined from domain knowledge is features themselves learned from a data-driven algorithm. While the great range both in feature choice and factors affecting performance makes data-driven learning of features a challenging task, the recent success of deep-learning techniques (i.e., spanning feature extraction and classification through layers of data-driven modeling) has raised new promise in this area. As would be expected, however, such approaches typically require substantially larger training datasets and incur greater computational complexity for inference, limiting their use in wearable and implantable systems today.

5.4.2 Inference

The objective of inference is to make decisions about data based on a previously constructed model. In the case of data-driven algorithms, the model is typically constructed from previously observed data. For instance, in the seizure-detection example of Figure 5.6, the model corresponds to a hyperplane in the feature space, dividing regions according to two classes we wish to distinguish. This gives us a decision boundary for performing inference by mapping data to the feature space via feature extraction.

As mentioned, data-driven methods have the benefit that they can enable high-order models for inference without requiring deep understanding of the underlying processes on a mechanistic level. This is beneficial in many application domains, including medical signal analysis, as such an understanding may be unavailable or too complex to yield viable models. On the other hand, machine-learning algorithms, combined with the availability of data, have led to unprecedented success of data-driven models, even in applications involving complex feature distributions of the signals. However, the feasibility of data-driven models is based on the assumption that the statistical distribution of data for inference is stationary or similar to the distribution of the previously observed training dataset. We note that, in the medical setting, many factors challenge this assumption; most notably, data typically acquired at one point in time and in a hospital or other in-patient setting may not be representative of broader scenarios affecting patient state, during which signal analysis is required in wearable and implantable devices. Protocols to address this challenge have been an active area of focus in the medical domain.

From the discussion above, it follows that data-driven inference requires two distinct phases (as shown in Figure 5.5): (i) model training, where a training dataset based on previously observed data is used to construct the inference model; and (ii) inference, where the constructed model is applied to new data to make decisions. Though model training is typically more energy intensive, particularly since it requires computations over a training dataset large enough to mitigate generalization error, inference is typically performed much more frequently. Thus, for real-time operation within wearable and implantable devices, inference is most commonly the primary focus. However, as will be discussed in the following sections, in many medical applications it is beneficial to localize various functions supporting or fully performing model training.

We point out that model training requires both a training dataset, but also labels specifying how elements of the training dataset correspond to the inference of interest. For instance, in the case of the seizure-detection example, the training dataset corresponds to segments of EEG while the labels correspond to annotations specifying which class (seizure or non-seizure) each segment belongs to. Generally, two approaches exist to model training: (i) supervised training (or supervised learning), where labels are provided explicitly; and (ii) unsupervised training (or unsupervised learning), where explicit labels are not provided, but rather labels are estimated using various algorithms. Both approaches have been employed in medical applications, though the typical protocol has been for clinical experts to provide some labeling to control model construction.

With regards to the models themselves, a great many options exist, particularly from machine learning, each presenting respective algorithms for training. Broadly, inference models can be classified as either *generative* or *discriminative*. Generative models aim to represent the joint distribution between a variable corresponding to the inference of interest and a variable corresponding to the data on which the inference is based. Discriminative models aim to represent the conditional distribution between a variable corresponding to the inference of interest and a variable corresponding to the data on which the inference is based. While generative models represent a more general relationship between the variables (e.g., the conditional distribution can be derived using Bayes rule), for typical inference tasks (such as classification of data) discriminative approaches provide a more directly relevant representation, typically leading to better performance in practical scenarios.

Among discriminative models, again, a great many options exist. Some of these, which have been researched for embedded inference in medical-sensor applications include support-vector machines (SVMs), linear regression, neural networks, and decision trees. Various implementations of these have been reported achieving low-energy and high-performance. In the sections that follow, the focus will be less on the specific choice of model and more on surveying practical issues related to energy-efficient realization, driven by the general structure of inference-system algorithms.

5.5 Specialized Inference Algorithms and Processors

Energy constraints challenge the ability to achieve advanced inference capabilities in wearable and implantable systems. Heterogeneous microprocessor architectures have provided one of the most effective paths to address energy efficiency in embedded computing systems. Heterogeneous architectures incorporate specialized hardware to substantially reduce computation-control overheads incurred, compared to microprocessors based only on fully programmable central-processing units (CPUs). As discussed below, inference systems for medical-sensor applications raise distinct opportunities for heterogeneous computing, thanks to their algorithmic structure. However, for broad usability, several factors must be considered.

5.5.1 Rationale

Although machine-learning algorithms have shown great promise for medical signal analysis, they can lead to excessive energy. Analysis of system energy shows that a primary challenge arises from the use of data-driven models in the inference stage. Specifically, as compared to analytical models, data-driven models typically do not have compact parametric representations. As a result, particularly when complex feature distributions are involved in a high-dimensional space, the models can be large. This results in significant computational and memory-accessing energy.

As an example, Figure 5.8 shows the performance and energy of a cardiac-arrhythmia detector based on ECG analysis, implemented using an SVM classifier [15]. In an SVM, the data-driven model is a set of

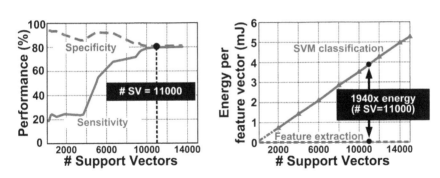

Figure 5.8 Performance and energy of cardiac-arrhythmia detector with respect to complexity of the data-driven inference model (i.e., number of support vectors) [15].

support vectors, which correspond to feature vectors selected from the training dataset, used to represent the decision boundary between data classes. Various approaches exist to reduce the number of support vectors, incurring some approximation of the decision boundary. The plot on the left shows the performance degradation that results, illustrating that for high performance a high-order model is required, particularly to address the complex feature distributions often encountered in medical applications. The plot on the right shows the SVM energy (for implementation on a low-power microprocessor), illustrating that such a high-order model causes the classification energy to dominate by orders of magnitude (i.e., the feature-extraction energy is also shown for reference.

The challenge observed in Figure 5.8 is seen frequently in medical-sensor applications, due to the complex feature distributions typically encountered. Fortunately, this challenge can be addressed through heterogeneous computing, thanks to the algorithmic structure of inference systems. In particular, as mentioned in Section 5.4.1, feature extraction requires a high level of programmability, due to the range of application signals and feature-performance considerations. However, especially in typical medical-sensor applications where the signals correspond to a limited number of low-sample rate time series, the computational complexity, and thus energy, is modest. Accordingly, feature computations can be delegated to a programmable CPU. On the other, inference (e.g., classification), which requires computation involving a potentially large data-driven model, is energy intensive, but is typically performed through specific computational kernels. Thus, it can readily be delegated to specialized hardware accelerators.

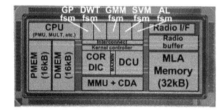

Figure 5.9   Die photo and measurement results of a heterogeneous microprocessor, demonstrating substantial energy savings thanks to hardware specialization [16].

Figure 5.9 shows a custom heterogeneous microprocessor for a range of medical detectors (including ECG-based cardiac-arrhythmia detection, EEG-based seizure detection, etc.) [16]. Here, the integrated CPU is used for programmable feature extraction, and configurable accelerator is used for low-energy inference. The table on the right shows the substantial reduction in overall energy achieved as a result of an architecture taking advantage of the algorithmic structure. However, the primary drawback of hardware specialization is the limited range of computations that can be supported. Thus, taking such an approach requires careful thinking of the configurability that must be supported within the hardware and exposed to application developers, in order to address the range of applications and aspects within these applications that are of interest. The following subsections explore this further.

5.5.2 Classifier Configurability

Though data-driven classification is typically performed through specific kernel functions associated with the data-driven models, in practice, the kernel functions present a range of parameter options. These options give rise to significant energy-performance tradeoffs, and it is natural with data-driven approaches that the optimal choice for the parameters is often driven by the characteristics of the application-level data. Thus, it becomes important to make these tradeoffs controllable at the application-design level, by selectively introducing configurability in the specialized hardware. Beyond such parameter configurability, the classifiers themselves may be used in specific ways within meta-algorithms. It is beneficial to also include support for common meta-algorithms, as any computations otherwise relying on the CPU can dominate energy. The following subsections examine various areas where configurability may be needed, and how this can be supported within accelerator microarchitectures.

5.5.2.1 Inference models

As discussed in Section 5.4.2, there are a great many different data-driven models that may be considered. While many of these will provide roughly similar functionality, they will vary in performance, depending on size of training dataset, training complexity, class imbalance, etc. Thus, in some cases it may be necessary to support different models. However, beyond this, there are specific models that may better align with the underlying processes of interest. As an example, dynamic time warping and/or hidden-Markov models may be amenable to processes characterized by specific evolution in state in a time dependent or time-independent manner. Thus, support in this direction may also be required.

Fortunately, many models, especially from machine learning, are based on some common principles, which lead to computational structure that can be exploited toward specialization. For instance, as discussed in Section 5.4.1, representation of data in a vector space is a common approach, since it enables similarity metrics based on geometric distance. This in turn leads to the computational structure of vector and matrix algebra. Though specialization in the form of vector-processor architectures has thus shown to substantially enhance the energy-efficiency of inference kernels, greater specialization, as well as optimization for low-power systems, have shown to take this further.

As an example, Figure 5.10 shows the structure of the microarchitecture employed within the inference accelerator, referred to as the machine-learning accelerator (MLA), of the microprocessor in Figure 5.9 [16]; the functional structure is shown on the left, while the high-level hardware

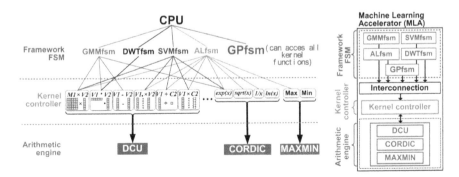

Figure 5.10 Processor architecture maps high-level inference computations to structured hardware [16].

structure is shown on the right. Within the accelerator, *Framework FSM* corresponds to a layer of finite-state machines (FSMs) that provide high-level data-flow and computation control for the various inference algorithms supported. Typically, such control would otherwise be provided by software executed on a microcontroller, at substantially higher energy cost. Dedicated FSMs are provided for several inference models and associated algorithms (as well as some common feature-extraction transformations), including Gaussian mixture models (GMMfsm), discrete wavelet transform (DWTfsm), support-vector machine (SVMfsm), and active learning (ALfsm). In addition, a general-purpose FSM (GPfsm) provides a high-degree of configurability, enabling user-defined high-level control. *Kernel Controller* corresponds to a layer of data-movement and data-structuring functionality for addressing the vector- and matrix-algebra computational structure of the models and associated algorithms. In addition to data structuring for linear algebra, this layer also performs data structuring for non-linear operations required within the inference algorithms (non-linear analytical functions, min/max search). For the specialized computations, this layer can be thought of as extending and playing the role of a direct-memory access (DMA) module, and estimated to play a critical role in enhancing energy efficiency by minimizing CPU intervention for data-flow control. Finally, *Arithmetic Engine* corresponds to a layer of computational functionality applied to low-level operands from the structures derived by the Kernel Controller. Given that such operations in fact constitute a small portion of the overall energy [17], the focus of hardware design at this layer is less on energy optimization, and more on supporting inputting of operands to optimize operation of the Kernel Controller. As an example, Figure 5.11 shows the microarchitecture of the Data Computation Unit (DCU). While the heart of the DCU is a multiply-accumulate (MAC) module, we see that various additional parameters require for kernel-function computations can be fed within the specialized two-stage pipeline. As will be described next, given the hardware-level support they require, the choice of parameters supported in this way is determined from application-level considerations.

5.5.2.2 Model parameters

In addition to supporting different types of data-driven inference models in order to address the range of processes of interest, most models also have a number of different parameters. We find that the precise choice of parameter values can strongly impact both performance and energy. The challenge when it comes to design of specialized processors is that with data-driven models,

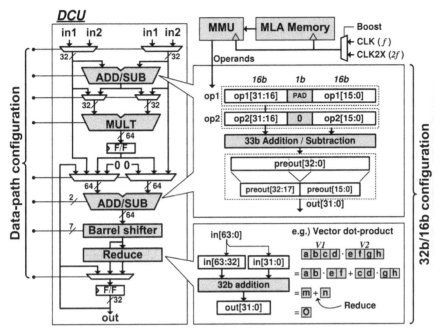

Figure 5.11 Microarchitecture of the specialized Data Computation Unit (DCU) [16].

the optimal choice of parameter values depends on the characteristics of application-level data. Thus, it is essential to determine which parameters and what range of values must be supported for control by application designers.

To provide an illustration, we focus on SVM classification, supported by the microprocessor of Figure 5.9. SVM classification is performed through a range of different *kernel functions*. These include linear, polynomial, and radial-basis-function (RBF) kernels, to name a few. These vary the ability for the model to be fit to feature-data distributions (other models have similar parameters, e.g., number of neurons and hidden layers in neural networks). Further, each of these kernel functions offers specific parameters (e.g., polynomial order in polynomial kernel, cost and gamma in RBF kernel). Further still, various algorithmic formulations of the kernels have been proposed, that raise additional tradeoffs. For instance, non-linear SVM kernels enhance the flexibility of classification models, but cause energy scaling with the number of support vectors. In Lee et al. [18] a formulation of second-order polynomial kernels is proposed, which employs a matrix representation of the support vectors to yield a linear computation.

This enables factorization of all support vectors into a single decision matrix, eliminating energy scaling with the number of support vectors. However matrix representation causes more severe energy scaling with the support-vector dimensionality (from linear to quadratic). Thus, the formulation is only beneficial if a large number of support vectors and relatively small feature-vector dimensionality is involved. Specifically, over 2000× energy reduction was achieved for a patient-generic ECG-based arrhythmia detector with over 15k support vectors and 21-dimensionality feature vector, while only 7.0–36× energy reduction was achieved for a patient-specific EEG-based seizure detector with 84–625 support vectors and 50-dimensionality feature vectors (after reduction from 432 dimensions).

Supporting the various kernel functions and kernel-function formulations in the microprocessor of Figure 5.9, a wide range in performance, energy, and required hardware resources is observed. Table 5.2 shows the measured performance, cycle count, and support-vector memory from the microprocessor for classification in a seizure-detector. Given the large range observed, making the tradeoffs accessible at the application-design level is essential.

5.5.2.3 Classification meta-algorithms

Beyond just binary classification, many applications need to support higher-level algorithmic functions, involving specific use of binary classifiers. Hardware-level support for these can again be beneficial by avoiding the energy overhead incurred through software implementation on a CPU. Two examples of such higher-level algorithms are ensemble classifiers and multi-class classifiers.

Ensemble classifiers may be used to construct a stronger classifier through the combination of multiple weaker classifiers. In machine learning, a strong classifier is one whose model can be trained to fit arbitrarily complex feature-data distributions, while a weak classifier is on whose model cannot be trained to fit arbitrary feature-data distributions, due to limitations in the set of all

Table 5.2   Tradeoffs for seizure detection with support vector machine [16]

	Seizure Detection				
	Performance			Cycle Count	SV Memory
Kernel	Sensitivity	Latency	Specificity	(kcycles)	Required (kB)
RBF	100%	4.8 sec	1.2/day	29.6	16.5
Poly (2nd order)	100%	4.4 sec	2.4/day	24.9	16.2
Poly Reform.				9.7	9.4
Linear	100%	15.0 sec	18.0/day	0.20	0.19

possible decision boundaries. Common categories of ensemble algorithms include boosting and bagging. For instance, in the popular boosting algorithm Adaptive Boosting (AdaBoost) [19], whose classifier structure is shown in Figure 5.12, weak classifiers are trained iterative and overall classifications are determined by weighted voting over all the iterations. During training, each iteration is biased to emphasize the importance of miss-fit instances from the training dataset, thereby resulting in perfect fitting over enough iterations. There are many reasons why an ensemble of weaker classifiers may be preferred over a single stronger classifier. For instance, the theory of AdaBoost suggests a strong classifier can be achieved with extremely weak classifiers (e.g., good performance is often achieved with single-node decision trees). This can yield overall reduced energy and/or hardware complexity, as well as various opportunities for energy-performance scalability through selection of weak-classifier iterations. Further, as will be discussed in Section 5.6, ensemble algorithms can enable various opportunities for addressing the resource requirements for model training.

Multi-class classifiers address the need for distinguishing data into multiple classes. Such problems are often encountered in medical-sensor applications where discrimination among multiple medical states may be of interest (e.g., identifying type of cardiac arrhythmia [20]). On the other hand, binary classification algorithms are much more prominent in machine learning. Thus, various meta-algorithms have been developed to address multi-class problems. A common approach taken is to employ multiple binary classifiers, and then define metrics based on their outputs, allowing for multi-class resolution. For instance, a set of one-vs.-all binary classifiers, also

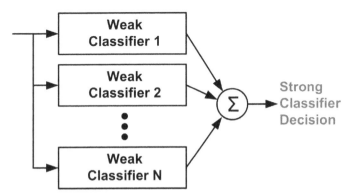

Figure 5.12   Ensemble learners construct a strong classifier with multiple weak classifiers.

providing a confidence metric, can be employed to determine the dominant result (this approach has been used with neural networks as well as classifiers such as SVMs, based on computing a marginal distance from a decision boundary). While generative classifiers may more naturally provide a confidence metric, in general a confidence metric may not be readily available from a discriminative classifier. Thus, an alternate metric often employed with binary, discriminative classifiers is based on all-vs.-all (AVA) voting. Figure 5.13 illustrates the approach, assuming resolution from 10 classes is desired. In AVA voting, a binary classifier is formed for every pair of classes, and the overall class is determined based on maximum voting over the outputs. As in the case of ensemble classifiers, hardware support for such mutli-class meta-algorithms can result in substantially better energy efficiency than software implementation via a CPU.

5.5.2.4 Active learning

Section 1.2 identified the need for patient-customization and dynamic adaptation of inference models. With data-driven models, medical sensors, capable of continuous or long-terms data acquisition from a specific patient, raise distinct opportunities in this regard. However, making data acquisition scalable is not enough. In particular, supervised machine-learning algorithms require a labeled training dataset. Current approaches involve annotation by clinical experts to provide labels for a training dataset. Even if medical sensors make data acquisition scalable, analysis of that data by clinical experts is not. *Active Learning* is an approach in machine learning that can help address this challenge. In active learning, data from an available pool is assessed

		Classes										
		0	1	2	3	4	5	6	7	8	9	
	0		o	o	o	o	o	o	o	o	o	9 ✓
	1	x		x	o	o	o	x	o	x	x	4
Classes	**2**	x	o		o	o	x	x	o	x	x	4
	3	x	x	x		x	x	x	o	x	x	1
	4	x	x	x	o		x	x	o	x	x	2
	5	x	x	o	o	o		x	o	x	x	4
	6	x	o	o	o	o	o		o	o	x	7
	7	x	x	x	x	x	x	x		x	x	0
	8	x	o	o	o	o	o	x	o		o	7
	9	x	o	o	o	o	o	o	o	x		7

Figure 5.13   10-class classification through all-versus-all voting on 45 binary classifier outputs.

to determine which instances would be most useful for a learning problem, such as inference-model training. In addition to reducing burden on clinical experts, performing such assessment on a medical-sensor device is valuable in order to mitigate communication energy (similar to embedded inference and feature extraction, as analyzed in Section 1.1). Accordingly, hardware support is beneficial to enhance energy efficiency compared to software implementation.

Figure 5.14 on the left shows a prototypical algorithm, illustrated for an ECG-based, patient-adaptive, cardiac-arrhythmia detector. First, a patient-generic seed model is constructed by employing training dataset based on population-level patient data. Then, along with on-going analysis for arrhythmia detection, the incoming data is assessed for inclusion in a new training dataset batch, based on a computed metric. When a defined number of data instances have been included in the batch, the batch is transmitted to a clinical expert (batch-wise transmission is performed to amortize wireless-communication overheads, such as transmitter-receiver synchronization). After analysis and model retraining, an update model is then uploaded to the device. This process can continue until a defined stopping criterion is met, and can be retriggered when a defined starting criterion is met. Figure 5.14 on the right shows the improvement in detection performance achieved by such an algorithm, as the detection model evolves from patient generic to patient specific.

The key aspect for enabling such an active-learning algorithm in an embedded device is continuous assessment of the sensed data by computing a metric for batch inclusion. A variety of such metrics have been proposed. For instance, for marginal-distance classifiers, such as SVMs, a natural metric is

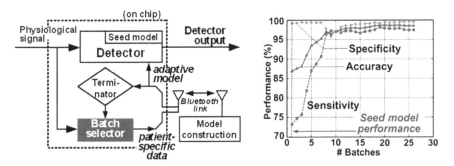

Figure 5.14 Prototypical patient-adaptive detector, illustrated through an ECG-based cardiac-arrhythmia detector [15].

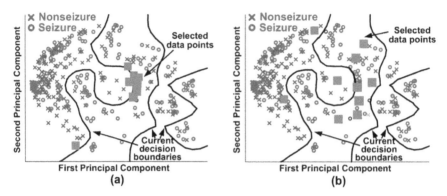

Figure 5.15 Batch data (a) with only marginal-distance metric, leading to concentrated selections, and (b) with both marginal-distance and diversity metrics, leading to coverage over large regions of the feature space [15].

the distance of a data instance from the current decision boundary. However, as a regularizer to achieve convergence with a reduced amount of data, diversity metrics have also been proposed specifically aim to cover larger regions of the feature space, rather than selecting instances for inclusion that are concentrated in a small region [21]. For illustration, Figure 5.15 shows the instances selected for a particular batch in a patient-adaptive EEG-based seizure detector, based on only a marginal-distance metric and based on both a marginal-distance and diversity metric. Such metrics have been particularly effective in cases where data is bursty in time, thus tending to be highly concentrated in a particular region during a given period. This is sometimes seen in medical-sensor applications, where patient data is correlated with activity states. The challenge that such metrics raise is that, in addition to computation of distance from the decision boundary, every instance of data must now also be involved in computation of distance from every other instance of data selected for batch inclusion, thereby substantially raising the computational complexity. Thus, hardware support for computing such metrics is beneficial, as was demonstrated in the heterogeneous microprocessor presented in Lee and Verma [16].

5.6 Training for Low-Energy Systems

While the previous section focused on energy-efficient classifier implementation, as discussed in Section 5.4 (and shown in Figure 5.5), a complete system also includes classifier-model training. Though classification is the primary focus for low-energy implementations, since it runs with high activity (possibly continuously) on a sensor device, training algorithms can also have important implications on classifier energy because of the models they result in. This section explores the opportunities raised during model training to enable low-energy classification.

5.6.1 Model Bit Precision

For embedded devices, computational precision has critical impacts on hardware resources (energy, memory size, area). This subsection look at how the tradeoff between classification accuracy and computational precision can be improved. For illustration, Figure 5.16 shows an example of a linear classifier $y = \text{sign}(\vec{w} \cdot \vec{x})$ in an image-detection system, where the bit precision of elements within the model $\vec{w}$ is quantized following double-precision training, as is typically done finite-precision implementations. As can be seen, the detection performance dramatically degrades with reducing bit precision below 8 bits.

The precision required strongly depends on the distribution of feature data with respect to the decision boundary, and the distribution of model parameter values themselves, which establish the decision boundary. While the data distribution is highly application dependent, unfortunately the parameter-value profile shown in Figure 5.17(a), corresponding to the linear-classifier weights for a cardiac-arrhythmia detector, is frequently encountered. Here, we see that a large number of the parameter values (i.e., elements of $\vec{w}$) have small value, while a small number of parameters values have much larger value. Thus, following quantization, many of the parameters are either zeroed or incur large percentage error, resulting in severe classification

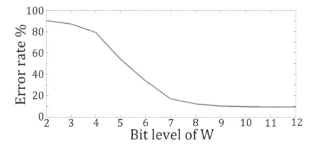

Figure 5.16 Classification error rate in image-detection system, where training is done with high precision, but derived model is then quantized [23].

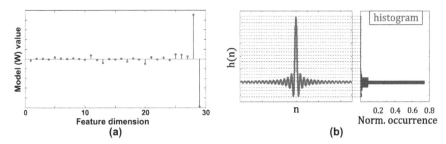

Figure 5.17 Profile of parameter values, tending to result in severe quantization errors, shown for (a) a linear classifier in arrhythmia detector, and (b) an FIR-filter feature-extractor in seizure detector [23].

error. In fact, a similar situation is commonly observed for finite-precision feature-extractions computations also. Figure 5.17(b) shows the FIR-filter coefficients (on the *left*) and a histogram of the coefficient values (on the *right*), for a band-pass filter used to extract spectral-energy features in an EEG-based seizure detector. Once again, a similar issue is observed. We briefly discuss two solutions to address this problem; the first optimizes the model-parameter values during model training to make them more conducive to quantization, while the second changes the finite-precision representation of the model-parameter values.

The first approach exploits the fact that model parameter values are often derived from optimization, aimed at minimizing the model's classification error over the training dataset. In such formulation, large-valued model parameters can be deliberately made less likely by more heavily penalizing them in the optimization cost function. For example, in the case of linear classifiers, one can use L2 regression (rather than L1 regression), which tends away from a small number of dominant parameter values in favor of more even distribution of parameter values. In addition to this, additional regularization terms can be added in the cost function to explicitly penalize large-valued parameters. A great deal of research has been done in this direction, with further details available in Tibshirani [22].

In addition to the way in which a chosen cost function imposes penalties during optimization, the training dataset can itself impact the distribution of model parameters. An issue that results commonly for linear classifiers is co-linearity of the training dataset, shown in Figure 5.18(a). Here, the training dataset labels (y in figure) change primarily due to one feature dimension (x_1 in figure), but the other feature (x_2 in figure) exhibits small changes,

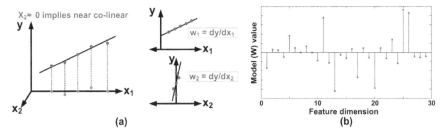

Figure 5.18   Co-linearity, (a) causing unreasonably large-valued model parameters, can be addressed through (b) principal-component regression, leading to a more robust distribution of model parameters.

having some linear correlation with variances in the first dimension. Since model parameters w_1 and w_2 represent the change in label values with respect to each dimension, respectively, the parameter associated with x_2 will have unreasonably large value. Such a condition can be easily checked by performing singular value decomposition (SVD) on the training dataset, represented as a matrix. Small singular values will indicate such co-linearity, and be handled by employing principal-component regression (PCR) rather than just linear regression. As an example, Figure 5.18(b) shows the model parameters derived through PCR instead of linear regression for the same linear-classification problem as Figure 5.17(a), achieving a more uniform distribution of parameters values and enabling a classifier more robust to quantization errors.

The second approach changes the finite-precision representation in order to minimize the quantization error incurred [23]. This approach starts by noting that the highly non-uniform parameter-value profiles in Figure 5.17 imply inefficient use of available dynamic range. Instead, consider using floating-point representation. Figure 5.19 compares the parameter value histograms for the FIR filter example, using 6 bits in a fixed-point representation versus 3 bits for the mantissa and 3 bits for the exponent in a floating-point representation. As seen in the parameter-value histogram, the floating-point representation results in more uniform distribution, and as seen in the quantization-error histogram (normalized to 1 LSB at the 6-bit level), it has greater density of error at values well below 1 LSB, while fixed-point representation has quantization error more evenly distributed up to 1 LSB. However, we also see that with floating-point representation, the quantization error can exceed 1 LSB (up to 7 LSB for 3-bit mantissa/exponent), while with fixed-point representation, the quantization error is limited to 1 LSB.

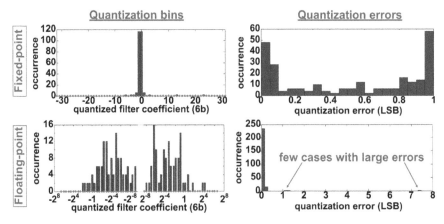

Figure 5.19   Comparison of quantization error for fixed-point (6 bit) and floating-point (3-bit mantissa, 3-bit exponent) representation of FIR filter coefficients [23].

To address this, floating-point representation raises another opportunity reduce error via a simple optimization discussed next.

The opportunity arises from the fact that in most systems, the outputs derived can often be scaled arbitrarily. Because of the substantially larger range enabled by a floating point operation, such scaling can readily be applied and can be used to minimize the quantization error of all parameters. The approach is shown in Figure 5.20, where the original parameters h_i (shown as red dots) are scaled by a factor α, which is chosen to yield new parameters (shown as blue squares) falling closer to the quantization levels. The aim is thus to find an optimal α that minimizes output error for an inner-product operation. This can be formulated into a well-defined optimization problem, and has shown to yield significant energy and accuracy advantages both for linear feature-extraction computations [23] and linear classification computations [23].

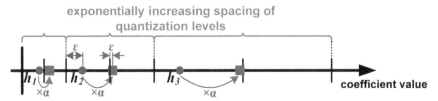

Figure 5.20   Quantization error minimization approach through parameter value scaling [23].

5.6.2 Dataset Imbalance

An issue that frequently arises in medical-sensor applications is training dataset imbalance. In particular, in many applications a lot of training instances may be available for a class corresponding to physiologic states while much more limited training instances may be available for pathophysiologic states. For instance, in the example of seizure detection, much more data is typically available for the non-seizure class compared to the seizure class. If the training dataset has such an imbalance, a model-training algorithm may very likely be biased towards the majority class.

While many model-training algorithms have included provisions for dataset imbalance, typically one of two approaches is taken. The first is to sample the training dataset selectively. This includes either down-sampling the majority class or up-sampling the minority class. A potential issue with down-sampling the majority class is that the reduced dataset may exhibit reduced statistical diversity and thus increase susceptibility to generalization error. However, methods for down-sampling have been proposed to mitigate this [24]. On the other hand up-sampling faces practical limitation. Namely, data from the minority class may be difficult to acquire or unavailable. Thus, the second approach is to either randomly duplicate training samples from the available minority-class data (this is also equivalent to adapting the penalty attributed for minority-class data within the optimization cost function), or to synthesize training dataset samples, based on the minority-class data distribution observed. Of course, both approaches have their limitations. Duplication may lead to models that over-fit to certain instances in the training dataset rather than enhancing statistical diversity, while data synthesis may result in statistics deviating from the true statistics.

It should also be noted that for imbalanced datasets, another issue is that conventional performance metrics such as accuracy and error rate may be of limited usefulness. For example, consider the case of Mammography cancer detection with 9,900 non-cancer images and 100 cancer images. A naive classification rule would be to classify everything as non-cancer, yielding a detection accuracy of 99%. Thus, generally in medical-sensor applications all four metric must be considered: true positive rate, true negative rate, false positive rate, and false negative rate.

5.6.3 Embedded Training

While much of the discussion thus far has focused on systems where inference is performed locally on a sensor platform, but training is done offline on a separate platform, applications exist where embedded training localized on the device may be desired. This is once again driven by the energy, bandwidth, and latency constraints of communication (since training is typically less frequent than inference, often bandwidth and latency constraints are more tolerable, thus often making training on a remote device viable). An example, where local training is believed to be advantageous is seizure prediction. Since seizure predication requires classification a long time in advance of seizure onset (tens of minutes), robust classification of inter-ictal data (i.e., between seizures) is necessary, elevating the importance of frequent patient customization to enhance specificity over the range of activities otherwise observed over a large population. The large amount of data that would need to be transmitted, especially in the case of many-channeled intra-cranial recording devices, can make the energy of communication to a separate platform prohibitive.

While the training dataset may be available within a sensing device, a challenge that arises with embedded training is how to acquire labels for training. Indeed, approaches for estimating training labels in an unsupervised or semi-supervised setting are an area of active research. However, medical applications can also raise specific opportunities. For instance, in the seizure-prediction application described above, a seizure detection, which is a simpler problem, can be employed to back annotate a seizure prediction training dataset.

On the systems level, the key challenge associated with embedded training is the computational resources it demands, particularly over the large training dataset typically required. However, because training is typically done infrequently and not under strict real-time constraints, many aspects, such as the energy required, can be amortized or do not require greatly increased computational throughput. However, one aspect that we note cannot be amortized is the amount of embedded memory required for the large training dataset. Recent research has begun to explore algorithms for embedded training that reduce the memory requirements. For instance, Wang et al. [25] exploit ensemble classification. Weak classifiers are chosen that show reduced susceptibility to generalization error, due to their limited flexibility in modeling a decision boundary. In a conventional boosting algorithm such as AdaBoost, a particular training dataset is used for all weak classifiers and is chosen to ensure low generalization error of the overall strong classifier that results. However, because AdaBoost trains each weak classifier iteratively, in fact the training dataset can be substantially reduced for each iteration, thanks to reduced susceptibility to generalization error, and a *new* training dataset can be acquired for each iteration to ensure adequate training dataset diversity over the entire ensemble. This allows the instantaneous memory requirements to be greatly reduced. Further, various approaches to active learning within a boosting setting have been proposed, that can be leveraged for selecting training dataset instances. For instance in FilterBoost [26], a metric is employed for selecting data instances from a pool to form the training dataset for each iteration. Figure 5.21 contrasts conventional AdaBoost with a training algorithm employing a selection metric modified from FilterBoost, showing a different and reduced training dataset for each weak-classifier training iteration. This algorithm, demonstrated over $65\times$ reduction in memory requirements, and over $10\times$ reduction in training energy for a seizure-detection system [25].

It is interesting to note that taking this approach, a tradeoff is introduced between the training complexity and the classification complexity. Using weaker classifiers within the ensemble is preferable for training because it enables greater reduction in the training dataset for each iteration; however, using stronger classifiers within the ensemble is preferable for classification because it enables training error convergence with fewer iterations. Figure 5.22 shows this tradeoff for the system in Wang et al. [25], where decision-tree weak classifiers of various nodes are employed.

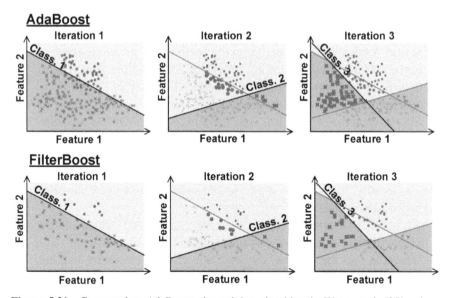

Figure 5.21 Compared to AdaBoost, the training algorithm in Wang et al. [25] reduces training dataset size by employing a new training dataset for each iteration.

Weak classifier	Stumps	4-node tree	7-node tree	
Iterations	64	22	19	Classifier complexity
NAND eq. gates	75k	27k	29k	
% fault affected	82%	87%	89%	
Trainer memory	7.2kB	10.4kB	13.6kB	Trainer complexity
Trainer clocks	0.33G	2.8G	2.6G	

Figure 5.22 Illustration of tradeoff between training complexity and classification complexity [25].

5.7 Summary

This chapter explored platforms and algorithms for mapping data-driven inference functions to low-power wearable and implantable medical devices. While such devices, due to their resource constraints, face significant challenges in realizing these functions, this chapter discussed the many opportunities for addressing the challenges on both the hardware and algorithmic levels. On the hardware level, the primary focus was on heterogeneous computational architectures, exploiting efficiency vs flexibility tradeoffs in computation, in the context of inference systems. A variety of design decisions were explored, related to designer configurability, generalization of microarchitectures across algorithms, and meta-algorithms. On the algorithmic level, the primary focus was on approaches to model training that reduce the computational-precision requirements, given that these raise important tradeoffs with energy. Approaches to embedded model training were also explored, especially focusing on how the amount of embedded memory required could be reduced.

While a number of promising directions were discussed, which indeed enable the possibility of bringing advanced algorithms into resource constrained wearable and implantable devices, this chapter closes by looking forward to the challenges of making such directions more broadly usable. Most notably the focus on heterogeneous architectures raises challenges with programmability. First, existing embedded software may not be able to directly exploit the efficiency gains provided by the architectures. Further, new software for doing this may in fact require in-depth knowledge of the hardware architecture, when in fact the clinical professionals responsible for deploying such devices are not likely to have advanced expertise in platform architectures. Second, the use of data-driven models for inference raise the opportunity to exploit existing and data and models, perhaps even through various forms of patient customization, in order to enhance patient specificity. However, currently methods for doing this are in a very early stage, making further research in this area necessary for substantially easing adoption in the medical domain. All of these challenges make it even more important for integrative collaborations between the communities of embedded device design, computational medical, and clinical practice. Such communities have grown increasingly accustomed to working with each other, rising significant promise for efforts forward.

References

[1] Csavoy, A., Molnar, G., and Denison, T. (2009). Creating support circuits for the nervous system: considerations for "brain machine" interfacing. *Proc. IEEE Symp. VLSI Circuits* 3, 4–7.

[2] Verma, N., Shoeb, A., Bohorquez, J., Dawson, J., Guttag, J., and Chandrakasan, A. P. (2010). A micropower EEG acquisition SoC with integrated feature extraction processor for a chronic seizure detection system. *IEEE J. Solid State Circuits* 45, 804–816.

[3] Mandal, S. and Sarpeshkar, R. (2008). Power-efficient impedance-modulation wireless data links for biomedical implants. *IEEE Trans. Biomed. Circuits Syst.* 2, 301–315.

[4] Shoeb, A., and Guttag, J. (2010). "Application of machine learning to epileptic seizure detection," in *27th International Conference on Machine Learning* (Sydney, NSW: ICML).

[5] Jang, K. J., Balakrishnan, G., Syed, Z., and Verma, N. (2011). Scalable customization of atrial fibrillation detection in cardiac monitoring devices: increasing detection accuracy through personalized monitoring in large patient populations. *Proc. IEEE Int. Conf. Eng. Med. Biol. Soc.* 2011, 2184–2187.

[6] MedlinePlus (2017). *Post Myocardial Infarction ECG Wave Tracings*. Available at: https://medlineplus.gov/ency/imagepages/18030.htm

[7] Nikus, K. et al. (2017). Updated electrocardiographic classification of acute coronary syndromes. *Curr. Cardiol. Rev.* 229–236.

[8] Heldt, T, Verghese, G. C., and Mark, R. G. (2013). "Mathematical modeling of physiological systems," in *Mathematical Modeling and Validation in Physiology: Applications to the Cardiovascular and Respiratory Systems*, eds J. J. Batzel, M. Bachar, F. Kappel (Berlin: Springer).

[9] Verma, N., Lee, K. H., Shoeb, A. (2011). Data-driven approaches for computation in intelligent biomedical devices: a case study of EEG monitoring for chronic seizure detection. *J. Low Power Electron. Appl.* 1, 150–174.

[10] Gotman, J., Ives, J., and Gloor, G. (1981). Frequency content of EEG and EMG at seizure onset: Possibility of removal of EMG artifact by digital filtering. *EEG Clin. Neurophysiol.* 52, 626–639.

[11] Bishop, C. (2006). *Pattern Recognition and Machine Learning*. Berlin: Springer.

[12] Khalid, S., Khalil, T., and Nasreen, S. (2014). "A survey of feature selection and feature extraction techniques in machine learning," in *Proceedings of the IEEE Science and Information Conference (SAI)* (Piscataway, NJ: IEEE).

[13] Guyon, I., and Elisseeff, A. (2003). An introduction to variable and feature selection. *J. Mach. Learn. Res.* 3, 1157–1182.

[14] Zijlman, M., Flanagan, D., and Gotman, J. (2002). Heart rate changes and ECG abnormalities during epileptic seizures: prevalence and definition of an objective clinical sign. *Epilepsia* 43, 847–854.

[15] Lee, K. H., and Verma, N. (2013). A low-power processor with configurable embedded machine-learning accelerators for high-order and adaptive analysis of medical-sensor signals. *J. Solid State Circuits* 48, 1625–1637.

[16] Lee, K. H., and Verma, N. (2013). A low-power microprocessor for data-driven analysis of analytically-intractable physiological signals in advanced medical sensors. *VLSI Symp. Circuits* 2013, C250–C251.

[17] Horowitz, M. (2014). "1.1 Computing's Energy Problem (and what we can do about it)," in *2014 IEEE International Solid-State Circuits Conference Digest of Technical Papers (ISSCC)*, Feb. 2014, pp. 10–14.

[18] Lee, K. H., Kung, S.-Y., and Verma, N. (2011). "Improving kernel-energy trade-offs for machine learning in Implantable and Wearable Biomedical Applications," in *Proceedings of the IEEE International Conference Acoustics, Speech and Signal Processing (ICASSP)*, Prague.

[19] Schapire, R. E., and Freund, Y. (2012). *Boosting: Foundations and Algorithms*. Cambridge, MA: MIT Press.

[20] de Chazal, P., O'Dwyer, M., and Reilly, R. B. (2004). Automatic classification of heartbeats using morphology and heartbeat interval features. *IEEE Tran. Biomed. Eng.* 51, 1196–1206.

[21] de Chazal, P., O'Dwyer, M., and Reilly, R. B. (2004). Automatic classification of heartbeats using morphology and heartbeat interval features. *IEEE Tran. Biomed. Eng.* 51, 1196–1206.

[22] Tibshirani, R. (1996). Regression shrinkage and selection via the lasso. *J. R. Stat. Soc. B Methodol.* 58, 267–288.

[23] Wang, Z., Zhang, J., and Verma, N. (2015). "Reducing quantization error in low-energy fir filter accelerators," in *Proceedings of the International Confrance on Acoustics, Speech and Signal Processing (ICASSP)*, Lujiazui.

[24] Japkowicz, N., and Stephen, S. (2002). The class imbalance problem: a systematic study. *Intell. Data Anal.* 6, 429–449.

[25] Wang, Z., Schapire, R., and Verma, N. (2015). Error adaptive classifier boosting (EACB): leveraging data-driven training towards hardware resilience for signal inference. *IEEE Trans. Circuits Syst.* 62, 1136–1145.

[26] Bradley, J. K., and Schapire, R. E. (2007). FilterBoost: regression and classification on large datasets. *Proc. Adv. Neural Inf. Process. Syst. Conf.* 2007, 185–192.

6

Wireless Power Transmission to mm-Sized Free-Floating Distributed Implants

S. Abdollah Mirbozorgi and Maysam Ghovanloo

Georgia Institute of Technology, US

Abstract

This chapter presents an inductive link for wireless power transmission (WPT) to mm-sized free-floating implants (FFIs) distributed in a large 3D space in the neural tissue. This WPT method is insensitive to the exact location of the receiver (Rx) as a design example for energizing such distributed implantable neural interfaces, which are found to be causing less damage to the surrounding neural tissue and lasting longer. The presented structure utilizes a high-Q resonator on the target wirelessly powered plane that encompasses multiple randomly positioned FFIs, all powered by a large external transmitter (Tx). Based on resonant WPT fundamentals, we have devised a detailed method for analysis, optimization of the FFIs, and explored design strategies including safety concerns, coil segmentation, and specific absorption rate (SAR) limits while using realistic finite element simulation models in HFSS with head tissue layers. We have built several FFI prototypes based on the optimization procedure to characterize their performance *in vitro*. Additionally, we provide a simple but accurate method for power transfer efficiency and power delivered to the load measurements with desired source and load resistances, which are matched with actual conditions that may be smaller or larger than 50 Ω, utilizing a network analyzer (NA) and a spectrum analyzer (SA) with 50 Ω ports.

6.1 Introduction

6.1.1 Free-Floating Implants (FFIs)

Brain machine interfaces (BMIs) are promising to restore sensory modalities and motor abilities in those with severe physical disabilities, such as blindness and paralysis [1–3]. For instance, by obtaining neural signals from a microelectrode array (MEA) placed in a small local population of motor cortex neurons, partial control of a robotic arm has been demonstrated in humans [4]. However, it is a known fact that most functions engage a large part of the brain, and functions such as precise control of limb movements are the results of complex activities among a large distributed network of neurons in different regions of the brain [5]. Therefore, a key portion of future BMIs require the ability to simultaneously interface with multiple neural sites distributed over a large area [6].

Current neural interfaces offer high density but localized coverage of the brain area in 2D or 3D [7–11]. However, the safety, biocompatibility, and longevity of the traditional highly integrated but tethered MEAs for clinical use is still one of the greatest challenges slowing down progress toward clinical use [12]. One of the key issues with these MEAs has been identified as the strain and micromotions resulted from the MEA being tethered to a central neural recording and/or stimulation hub, causing damage in the surrounding neural tissue, and blood–brain barrier (BBB) [12, 13]. Neuroscientists and clinicians are now suggesting that advanced BMIs capable of wireless neural recording and stimulation while being small enough to be distributed over a large area of the brain, and float freely with the surrounding brain tissue without any tethers to a central hub would stand a better chance of operating over many years and decades [6, 14–16].

In the distributed BMI architecture, implants are very small, independent from one another, free-floating, and truly wireless [17]. Two-key constrains for these new free-floating distributed implants (FFIs) are size and sufficient wireless power to support a meaningful level of functionality. While the system-level design of the FFIs for a specific application is out of the scope of this article, we are focusing on the key aspect of delivering sufficient and robust (i.e., insensitive to misalignment) power to the load (PDL) in a few mW range to a generic FFI-based BMI, despite extreme size constraints (1 mm^3) at the highest possible power transfer efficiency (PTE), without violating the specific absorption rate (SAR) limits. This is a common requirement among many FFI architectures that should facilitate their design and development for various BMI applications.

Considering potential benefits of the FFIs in terms of longevity and biocompatibility, several groups have started investigating the feasibility of various methods to fabricate and wirelessly operate these devices [14, 16]. Energy harvesting from glucose fuel cell, thermoelectric, or piezo transducers have been proposed, but unlikely to provide sufficient power for the desired functionality [18]. Ultrasound, optical, and high-frequency electromagnetic field are viable solutions, currently being investigated as potential means to energize the FFIs [14, 19, 20]. Ultrasound, however, cannot penetrate through the skull, limiting its usage to peripheral nerve interfacing or from an implanted device that receives power through an inductive link, and relays it over a short distance across a few membranes and cortical tissue [14]. Moreover, they are quite sensitive to transducer alignment under a narrowly focused ultrasound beam, requiring an array of Txs for beamforming, which would multiply the amount of required power by the number of Txs [21]. Resonance-based near-field electromagnetic wireless power transmission (WPT) is well-established and demonstrated the ability to provide high-power density at extended range [22]. Even though several focused high-frequency WPT links for a single miniature implant have been proposed [14, 19, 20], there has been little information on a design guideline for inductive power transmission to multiple small randomly distributed FFIs with arbitrary misalignments over a large area of the brain [15, 23]. It is not yet clear which one of these methods will dominate this particular arrangement and lead to the most robust and safe method for WPT and communication with small electronic devices implanted deep in the body [14, 19, 20, 24].

6.1.2 WPT Measurement Accuracy

Wireless power transfer technology has been around for over a century [25, 26]. In recent years, WPT is made even more viable by better implementations and advances in technology that utilize WPT [22]. However, WPT still has considerable room to improve and vast potential to become ubiquitous in many fields from medicine to consumer electronics and transportation. These various applications have different characteristics, such as power levels, which change their design requirements [16, 26, 27]. The source and load impedances are among parameters that significantly influence the performance of the inductive links by changing both PDL and PTE [28–30]. Therefore, these impedance values are important parameters in the design of inductive links for different applications from μW to kW range [16, 26].

Conventionally, several parameters of an inductive link, including coils' inductances, parasitic resistances, and mutual coupling between each pair of coils in the link, need to be measured to calculate the PTE and PDL using equations in Ahn and Ghovanloo [16, 28–30]. These models are often simplified to make it easy to present and understand the design and optimization procedure by ignoring the small couplings in the multi-coil inductive links. The maintain high accuracy for these measurements is difficult because:

(i) there are numerous parameters that need to be measured in the presence of parasitics imposed by the measurement setup, (ii) parameters are inter-related and each parameter must be measured when the coils are located at identical positions in which the other parameters are measured, and (iii) sometimes small misalignments among the coils are unavoidable. Moreover, coil couplings are often measured with instruments that have 50 Ω ports, but the measured value would be slightly different if a port with different impedance is used. These limitations reduce the accuracy of the measured results, consequently affecting the optimal designs, especially for applications sensitive to coil movements, limiting accurate measurements to simple inductive links with two coils. Even with two-coil links, to measure PTE, one should measure transmission coefficient (S21), which does not eliminate any parameters, however, it is only valid for 50 Ω source and load resistors.

6.2 Modeling and Analysis of Inductive Links

One of the key challenges in design of FFIs is the weak coupling between the embedded miniature Rx coil and the external Tx coil at distances that are much larger than the Rx coil diameter, which makes it difficult to achieve sufficient PDL through a two-coil link [16]. In 2007, a group of physicists at MIT came up with coupled-mode magnetic resonance theory (CMT) to wirelessly transfer power across a long distance at high PTE, using four optimized coils, two of which were high-Q LC-tank resonators [22]. This method shows its benefits in a variety of media and high-power applications, such as charging laptops and electric vehicles [31]. However, its usage in transcutaneous WPT to ordinary implants was limited because, as explained in Kiani and Ghovanloo [30], the resonators do not offer any significant advantage when the Tx–Rx separation is relatively small.

The two-coil and three-coil inductive links circuit models are presented in Figures 6.1(a) and (b), respectively.

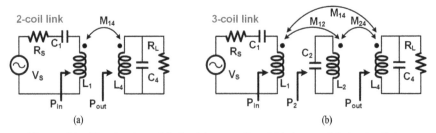

Figure 6.1   The circuit models for (a) two-coil and (b) three-coil inductive links.

We present a three-coil inductive link arrangement for powering distributed implants using an additional resonator at the Rx plane. We analyze and compare the two-coil and three-coil links by targeting the FFI application in this chapter. We have found that when the size of the Rx coil is very small, using the high-Q resonator can significantly improve the PTE, coverage area, and robustness against misalignments because of the magnetic field boosting and homogenizing effect over the area encompassed by the resonator, particularly when the resonator is segmented. The added resonator significantly extends the coverage area of the Tx coil, where the Rx can receive >1 mW without surpassing the SAR limit. Since the proposed resonator is more strongly coupled with the Tx coil, L_1, than the Rx coil, L_4, we have considered it as a Tx resonator, and refer to it as L_2.

Generally speaking, the resonator plays the role of a matching circuit to separate the load and source impedances from the inductive link, which result in the PTE improvement [29].

Figures 6.2(a) and (b) present the equivalent circuit models of the 2-coil and the 3-coil inductive links, respectively, and the M_{ij} (i and $j = 1, 2, 4$) is the mutual coupling between coils L_i and L_j. To explain the PTE improvement, we conventionally use the basic PTE Equations (6.1) and (6.2), presented for the two-coil and three-coil links, respectively.

$$\text{PTE}_{2-\text{coil}} = \frac{k_{14}^2 Q_1 Q_{4L}}{1 + k_{14}^2 Q_1 Q_{4L}} \times \frac{Q_{4L}}{Q_L}. \tag{6.1}$$

$$\text{PTE}_{3-\text{coil}} = \frac{k_{12}^2 Q_1 Q_2}{1 + k_{12}^2 Q_1 Q_2 + k_{24}^2 Q_2 Q_{4L}} \times \frac{k_{24}^2 Q_2 Q_{4L}}{1 + k_{24}^2 Q_2 Q_{4L}} \times \frac{Q_{4L}}{Q_L}, \tag{6.2}$$

where k_{ij} is the coupling coefficient between L_i and L_j, while $k_{ij} = M_{ij}/\sqrt{L_i L_j}$, and Q_i and Q_{4L} are the quality factor of the coil L_i and load quality factor, respectively. In Equation (6.2), M_{14} is ignored for simplification since it is smaller than M_{12} and M_{24} (M_{14} is shorted in the

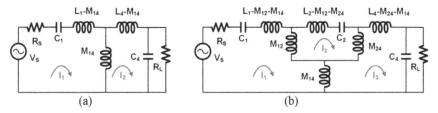

Figure 6.2 The equivalent circuit models of (a) two-coil and (b) three-coil inductive links.

Figure 6.2(b)). For target application of FFI, (i) k_{24} in Equation (6.2) > k_{12} in Equation (6.1), (ii) k_{12} in Equation (6.2) $\gg$ k_{12} in (6.1) and k_{24} in (6.2), and (iii) Q_2 (resonator) > Q_1 (Tx) and Q_4 (Rx), which are reduced by the source and load resistors. Therefore, the PTE of the three-coil link is much higher than the two-coil link even by ignoring k_{14}.

To accurately calculate the PTE of the three-coil link, we use the equivalent circuit model of Figure 6.2(b) and consider k_{14} in our analysis. Based on the power flow from the source to the load, we can define the efficiencies between the elements as provided in Figure 6.1(b). For the three-coil link, the efficiency equals $\text{PTE}_{3-\text{coil}} = P_{\text{out}}/P_{\text{in}}$. If η_{14} is the PTE between L_1 (Tx) and L_4 (Rx) coils, superposition suggests,

$$P_{\text{out}} = P_{\text{in}} \times \eta_{14} + P_2 \times \eta_{24}. \tag{6.3}$$

where $P_2 = P_{\text{in}} \times \eta_{12}$, η_{12} is the efficiency between L_1 and L_2 (resonator), and the η_{24} is the efficiency between resonator (L_2) and Rx (L_4). Then from [29],

$$
\begin{aligned}
\text{PTE}_{3-\text{coil}} &= \eta_{14} + \eta_{12} \times \eta_{24} \\
&= \frac{k_{14}^2 Q_1 Q_{4L}}{1 + k_{14}^2 Q_1 Q_{4L}} \times \frac{Q_{4L}}{Q_L} + \frac{k_{12}^2 Q_1 Q_2}{1 + k_{12}^2 Q_1 Q_2 + k_{24}^2 Q_2 Q_{4L}} \\
&\quad \times \frac{k_{24}^2 Q_2 Q_{4L}}{1 + k_{24}^2 Q_2 Q_{4L}} \times \frac{Q_{4L}}{Q_L}.
\end{aligned}
\tag{6.4}
$$

η_{12} is quite high since M_{12} between the Tx coil and resonator is large, and $\eta_{24} > \eta_{14}$ since L_2 is in the same plane as L_4. Additionally, L_2 is not influenced by R_S. In the three-coil structure, the contribution of the L_1 in delivering power directly to the Rx is similar to the two-coil link, which is the first term in Equations (6.3) and (6.4). The resonator, adds the second term in these equations, which can be considerably higher than the first term because $\eta_{12} \gg \eta_{14}$ and $\eta_{24} > \eta_{14}$. To maximize the PTE of the link, we have determined the optimal parameters of the coils and operating frequency utilizing the coil design theory in Kurs et al. [22, 29, 30, 32], and swept key parameters of the three-coil link model in simulation for fine tuning.

In cases where the Tx and Rx have almost the same size and the distance between them is roughly 10 times smaller than the coils' outer diameters (d_{out}), then there is no need to use resonators, and two-coil link would be sufficient. However, the coil arrangement in this article is quite different. If one intends to extend the coil distance and powering a coil with significant size difference between Rx and Tx coils, then using resonator would significantly improves the PTE. On the other hand, the resonator in our design significantly increases the coverage area that Rx devices can be located, while in previous design of wireless links for powering mm-sized coil, the EM-field is focalized to the center which is the only valid location for the only Rx device.

6.3 Design Procedure

We present an inductive link for WPT to mm-sized FFIs distributed in a large 3D space in the neural tissue that is insensitive to the exact location of the Rx. The presented structure utilizes a high-Q resonator on the target wirelessly-powered plane that encompasses multiple randomly positioned FFIs, all powered by a large external Tx. Based on resonant WPT fundamentals, we have devised a detailed method for optimization of the FFIs and explored design strategies and safety concerns, such as coil segmentation and SAR limits using realistic finite element simulation models in HFSS including head tissue layers, respectively. We have built several FFI prototypes to conduct accurate measurements and to characterize the performance of the presented WPT method.

The exemplar FFI considered for the presented coil inductive link design and optimization, is a small implant (<0.001 cc) in the form of a free-floating, untethered pushpin with a small footprint (1×1 mm^2) on the brain that will be placed by the neuroscientist or neurosurgeon over the cortical surface in the areas of interest, as shown in Figure 6.3 [15]. In this approach, the cortical area of interest, where the FFIs are located, is enclosed by a lightweight, flexible, hermetically sealed and passive LC resonator with high Q-factor.

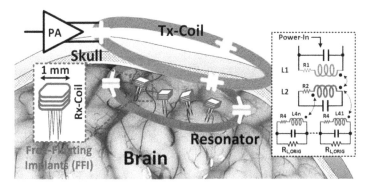

Figure 6.3 Conceptual rendering of the proposed WPT link for distributed free-floating implants (FFI) encompassed by an implanted high-Q resonator. Inset: simplified equivalent circuit model of the proposed inductive link.

In larger hosts, such as humans and non-human primates (NHPs), FFIs will be powered and interrogated by an external transceiver that is worn on top of the head in the form of a cap, while in smaller hosts, such as rodents, the transceiver can be combined within the experimental arena or the homecage [33, 34]. Here, we presents a detailed method for designing the presented WPT link, considering the properties of various tissue layers in the human head, including skin, fat, bone, dura, and brain.

In the following sub-sections, we present a design procedure for the presented three-coil inductive link to power up mm-sized distributed FFIs. It includes full-wave 3-D electromagnetic field models, constructed in the HFSS (ANSYS, Cecil Township, PA), to optimize the three-coil inductive link, including SAR assessment.

6.3.1 System Configuration and HFSS Simulation

In the presented three-coil transcutaneous WPT link for FFIs, the Tx coil is placed above the head, while the high-Q resonator and mm-sized Rx coils are placed in the same plane on the surface of the brain, as shown in Figures 6.4(a) and (b). This scalable configuration allows for wireless powering of a large number of FFIs with arbitrary positions and orientations across a large area encompassed by the high-Q resonator, thanks to the three-phase excitation scheme in [15, 35]. Figure 6.4(b) shows the cross-section of the presented link, considering various tissue layers and depths of various components. In our experimental setup, the Tx (L_1) and resonator (L_2) coils were insulated from the surrounding tissue by Polyimide (Kapton) films, while the Rx coil (L_4) was coated with PDMS.

The inductive link and surrounding medium, including the human head tissue layers, skin, fat, bone, CSF, dura, and brain, are simulated in HFSS according to Figure 6.4(b) model. Figure 6.4(c) compares the Poynting vectors (real part of power density) of the conventional two-coil inductive link and the presented three-coil inductive link with identical dimensions. The red area indicates a high directional flux density, required for high rate energy transfer per unit area. It can be seen that the red area in the presented three-coil structure is considerably wider at the Rx location compared to the two-coil link because of the magnetic field boosting and focusing effects of the L_2 resonator in the region of interest. The purpose of using resonator in this system is not only for improving PTE. We extended the area of coverage by adding the resonator. So, the area of coverage (valid location for Rx

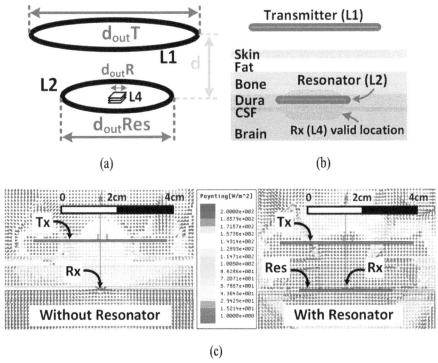

Figure 6.4 (a) The presented 3-coil link configuration for FFI using a high-Q resonator (d_{outRes}: diameter of the resonator) around a d_{outR} = 1 mm Rx coil; (b) Cross-section of the transcutaneous WPT link including tissue layers, location of various components, and coil insulation, and (c) HFSS simulations comparing the Poynting vector without (*left*) and with (*right*) a passive high-Q resonator.

devices) is one input parameter for optimizing the presented three-coil link, defined by user (application requirements).

6.3.2 Inductive Link Design Rules

To design the presented three-coil link for powering up FFIs with mm-sized Rx coils, distributed within the high-Q resonator, we have devised the algorithm in Figure 6.5. The flowchart includes four main sections: (i) input parameters, (ii) L_4 optimization based on previous work in Ahn and Ghovanloo [16], (iii) optimization of L_1 and L_2, and (iv) output parameters. The input parameters, which are defined by the features and requirements of the application and coil fabrication constrains, are (i) FFI geometry,

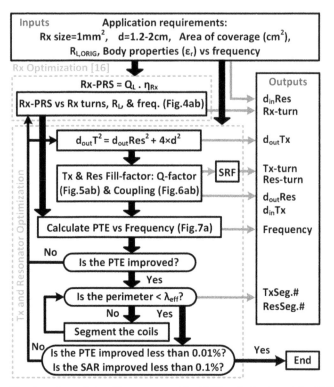

Figure 6.5 The flowchart of algorithm presented for PTE optimization of the FFI 3-coil link considering body layers and SAR limits.

(ii) thickness of various tissue layers, (iii) area of coverage, i.e., valid locations for placement of the FFIs, and (iv) R_{L}, which is calculated based on the FFI power and voltage requirements, in this case, 1 mW and 1 V across the Rx coil, respectively. The coverage area is determined by the resonator inner diameter, $d_{\mathrm{in}}\mathrm{Res}$.

Key properties of Rx are defined in Ahn and Ghovanloo [16] with a parameter known as the Rx power reception susceptibility (Rx-PRS), which indicates how efficiently Rx can receive power under a given magnetic field exposure,

$$Rx - \mathrm{PRS} = Q_{\mathrm{L}} \times \eta_x, \tag{6.5}$$

where $Q_{\mathrm{L}} = \omega L_4/(R_4 + R_{\mathrm{L}})$ is the loaded Q-factor of the Rx coil, $\eta_{\mathrm{Rx}} = R_{\mathrm{L}}/(R_4 + R_{\mathrm{L}})$ is the Rx internal efficiency, and $R_{\mathrm{L}} = (\omega L_4)2/R_{\mathrm{L,ORIG}}$ is the parallel to series RLC transformed version of the actual load resistance, $R_{\mathrm{L,ORIG}}$, assuming it has a large value, i.e., low-power consumption [16].

To achieve higher PTE and PDL on the Rx side, the Rx coil number of turns (Rx-turn) and loading (R_L) should be adjusted to achieve higher Rx-PRS. To have 1 mW of received power at 1 V across the Rx coil, the equivalent loading is $R_{L,ORIG}$ = 500 Ω.

To optimize the Rx coil using Equation (6.5), we modeled the Rx and head tissue layers (Figure 6.4(b)) in HFSS. The Rx coil has a square shape with a 1-mm inner side length to be wound around a 1 mm × 1 mm silicon die. The outer side length of Rx implants should not be more than 1.3 mm including the PDMS coating. Therefore, the Rx's wire diameter is chosen 100 μm, or 38 American wire gauge (AWG), and to complete the HFSS model, 50 μm PDMS isolating protection layer is added all around the Rx. The pitch between tightly wound turns is considered 120 μm, allocating 10 μm for wire insulation.

This model is used to optimize the Rx coil considering head tissue layers. Figure 6.6(a) presents the HFSS simulation results of the Rx-PSR as a function of frequency and Rx-turn. When the Rx coil is placed in tissue, Q_4 decreases significantly and its peak shifts to the lower frequencies. Therefore, it is important to consider the effect of surrounding tissue properties in the design and chose the carrier frequency, f, accordingly. As a result, by increasing the Rx-turns, Rx-PRS reduces and the peak frequency shifts to even lower frequencies.

As shown in Figure 6.6(b), heavier loading, i.e., lower R_L, also shifts the Rx-PRS peak to lower frequencies. Although in this paradigm, the design of Rx is considered independent of Tx because of the fact that the Rx size is very small and EM-filed is uniform around it [16], selected f should represent

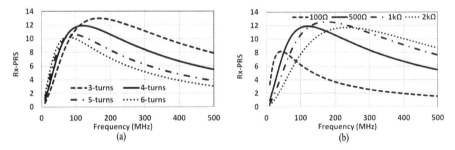

Figure 6.6 (a) The HFSS simulation of Rx efficiency (Rx-PSR) as a function of frequency for different Rx turns, with the presence of body layers and $R_{L,ORIG}$ of 500 Ω, and (b) the HFSS simulation of Rx efficiency (Rx-PSR) as a function of frequency and actual load resistor ($R_{L,ORIG}$).

a balance between the most efficient operating parameters of both Rx (Rx-turns and R_L) and Tx coils (L_1 and L_2). Therefore, the Tx design and its characteristics as a function of frequency should be considered for choosing the operating frequency and maximize the PTE of the entire inductive link for the FFI.

We have simulated the Tx coil, L_1, and resonator, L_2, in HFSS using the same setup as the Rx coil. Both Tx and resonator are to be made of copper foil with a thickness of 100 μm, covered by 25 μm thick polyimide (Kapton) film with relative permittivity, ε_r = 3.4. The distance from L_1 to both L_2 and L_4, d, depends on the anatomical position of the FFIs and the species. In humans and NHP, d can be chosen anywhere between 1 and 2 cm. As initial values, we considered the thickness of skin, fat, and bone to be 1 cm plus a 0.2-cm air gap between L_1 and the outer surface of skin. To have 7 cm^2 of brain coverage, the inner diameter of the resonator, d_{in}Res = 3 cm. To find the optimum width for L_2, the algorithm sweeps the resonator outer diameter in HFSS. As shown in Figure 6.7(a) inset, Q_2 also decreases significantly in the presence of tissue layers. These HFSS simulation results show that smaller widths of the implanted coils provide higher Q-factor at higher frequencies, in lines with the findings in Kiani and Ghovanloo [30]. Therefore, we chose the resonator width to be 1 mm, which corresponds to an outer diameter of 3.2 cm for the circular copper foil resonator.

Next, we determine the Tx coil outer diameter, $d_{out}T$, that maximizes the coupling coefficient between L_1 and L_2. By selecting d = 1.6 cm, and calculating $d_{out}T$ from [36, 37],

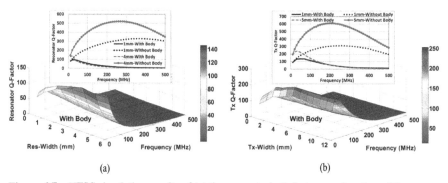

(a)   (b)

Figure 6.7 HFSS simulation results of (a) the resonator's Q-factor as a function of frequency and width of the resonator, and b) the Tx's Q-factor as a function of frequency and width of the Tx coil.

$$d_{\text{out}}T^2 = d_{\text{out}}\text{Res}^2 + 4 \times d^2 = 4.5 \text{ cm} \qquad (6.6)$$

The inner diameter of L_1 is swept, and simulated Q_1 is shown in Figure 6.7(b) as a function of the frequency and the width. The inner diameter of L_1 is chosen to be 3.5 cm as the smallest size in Figure 6.7(b) that offers a value close to the highest Q_1. Similar to the other two coils, Q_1 decreases significantly in the presence of tissue, resulting in the choice of lower carrier frequency and conductor width.

In addition to the Q-factor, coupling coefficients between three coils play a key role in PTE optimization. Here, we only consider k_{12}, between $L_1 - L_2$, because k_{24} and k_{14} are quite small due to the small size of L_4. Figures 6.8(a) and (b) show k_{12} as a function of the frequency and L_2 and L_1 widths, respectively. To determine the optimal f for the wireless link that maximizes the PTE, we followed the design procedure in Figure 6.5 flowchart, while considering the Tx-resonator's Q-factor, k_{12}, and Rx-PRS as the inputs, derived from the HFSS simulation results and coil design theory [29, 30].

Figure 6.9(a) shows the PTE variations in HFSS as f is swept from 25–200 MHz with various resonator widths. It can be seen that $f = 60$ MHz is the optimal choice for WPT in our target application, considering tissue properties of the human head, which permitivities are shown in Figure 6.9(b). By choosing $f = 60$ MHz, the optimum Rx-turns and R_L, can be found to be four turns and 500 Ω, respectively. As a rule of thumb, to ensure the self-resonance frequencies (SRF) of all the coils are at least four times higher than f, we minimized the number of turns for L_1 and L_2 to only one turn. The optimal number of turns for L_4, however, is found according to Figure 6.6 at 60 MHz. Expectedly, both SRFs and Q-factors of all coils significantly decrease in the presence of tissue layers.

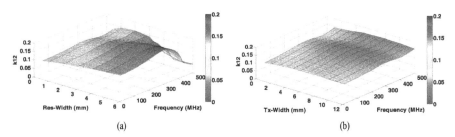

(a)   (b)

Figure 6.8   HFSS simulation results for coupling coefficient between L_1 and L_2 (k_{12}) as a function of frequency and (a) width of L_2, and (b) width of L_1.

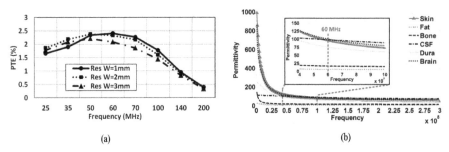

Figure 6.9   (a) HFSS simulation of the PTE for the presented link as a function of frequency and resonator width. (b) Relative permittivity of human head issue layers for skin, fat, bone, dura, and brain vs. frequency [38].

6.3.3 Coil Segmentation

The last steps in the presented optimization algorithm are segmentation and SAR limit verification. A radius of $\lambda/2\pi$ from the center of the conductive loop is generally considered the near-field domain of the loop, where λ is the operating wavelength [39]. Electrically small loops, such as a one-turn coil with a perimeter of less than $\lambda/10$ can produce strong and uniform magnetic fields in the near field [40]. To maintain perimeter of the loop to be less than $\lambda/10$ when the operating frequency increases, the loop should be segmented. Segmenting the loop prevents phase-inversion and nulls in the current distribution along the loop that cause electromagnetic field cancellation at the location of the Rx [40]. Therefore, it improves the uniformity of the current distribution, SAR, and inductive coupling. Wavelength λ can be calculated from,

$$\lambda = \frac{C}{f \times \sqrt{\varepsilon_r}}, \tag{6.7}$$

where C is the speed of light in vacuum (3×10^8 m/s), f is the operating frequency, and ε_r is the relative permittivity of the surrounding environment, i.e., the tissue layers in this case. Electrical properties of human tissues besides ε_r are conductivity, σ_{eff}, which controls the propagation, reflection, attenuation, and other behaviors of electromagnetic fields in the body [41]. These properties depend on the type of tissue and the frequency of operation. Utilizing the lowest frequency results in less attenuation as long as it does not limit the Rx coil Q-factor. Since human body is weakly magnetic, its permeability, $\mu_r \approx 1$. The maximum value of ε_r for tissue layers in human head at 60 MHz equals 100. Using Equation (6.7), $\lambda = 50$ cm. The perimeters of

L_1 and L_2 are 14.14 cm and 10 cm, respectively, and the length of each segment must be less than $\lambda/10 = 5$ cm. Thus, we need to segment L_1 and L_2 by 3 and 2, respectively.

6.3.4 SAR Calculations

Figure 6.10(a) compares the electrical field (E-field) in the resonator plane for the complete loop vs. segmented Tx coil and resonator, comparing uniformity of current distribution. Since SAR $= \sigma|E|2/\rho$, where σ and ρ are the tissue

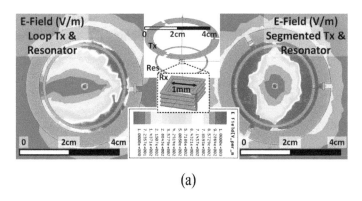

(a)

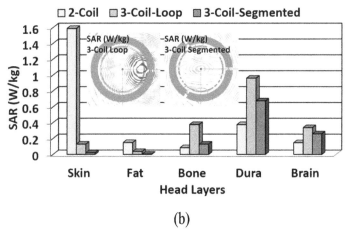

(b)

Figure 6.10 (a) HFSS simulation of the E-field, comparing the complete loop and segmented coils, showing uniformity of the current distribution in the case of three-coil link, (b) HFSS simulation results for the SAR simulation presenting the peak of average SAR values for tissue layers in conventional two-coil link, three-coil link with loop coils, and three-coil link with segmented coils, while delivering identical power of 1.3 mW to the receiver.

conductivity and density, respectively, by improving the uniformity in the E-field, both SAR and PTE would improve, by increasing power transmission under SAR limitation of 1.6 W/kg [42].

Figure 6.10(b) presents the HFSS simulation of SAR for the tissue layers of the human head when the tissue mass is considered 1 g and the input power levels are set to deliver an identical amount of power (1.3 mW) through 2-coil, 3-coil with complete loop resonator, and 3-coil with segmented resonator to the Rx. In this simulation, we have considered the average SAR of the layers with maximum absorption, which are skin and dura for two-coil and three-coil links, respectively. The results show that SAR of the three-coil link with segmented coils is less than that with complete loop coils, which translates to allowing transmission of more power, while improving the uniformity of the E-field.

6.4 Experimental Results

6.4.1 Implementation and Setup

We implemented the coils and characterized the performance of the presented three-coil inductive link at 60 MHz. Using a Vector Network Analyzer (VNA), we measured the specifications of the presented inductive link and summarized them in Table 6.1. The Tx and resonator coils are one turn planar made of flexible copper foil with 0.1 mm thickness, weight of 0.09 g/cm^2, and conductivity of 0.586 MΩ/cm (C110 Copper Foil, Online Metals, Seattle, WA, USA). The Rx coil is a four-turn wire-wounded coil, implemented

Table 6.1   Specifications of the inductive link at 60 MHz with/without the Surrounding Tissue

Parameter	L_1 (Tx)	L_2 (Res.)	L_4 (Rx)
Inductance, L (nH)	71.5/71.3	68.8/67.7	32.5/31
Quality Factor (Q)	210/570	62/280	29/31
Outer diameter, d_o (mm)	45	32	1.2
Inner diameter, d_i (mm)	35	30	1
Conductor width, W (mm)	5	1	0.1
Conductor thickness, t (mm)	0.1	0.1	0.1
Number of turns (N)	1	1	4
SRF (MHz)	310/570	300/720	580/925
Type of coil	Planar	Planar	38 AWG
Coupling coefficient	k_{12}	k_{24}	k_{14}
	0.132/0.135	0.0076/0.0076	0.002/0.0023

with 38 AWG wire and 10 μm thick insulating polyamide coating. Figures 6.11(a–c) show the Q-factor of the coils with and without the presence of head tissue layers for Rx, resonator, and Tx coils, respectively. It can be seen that the Q-factors of all coils are significantly reduced because of the surrounding tissue layer properties.

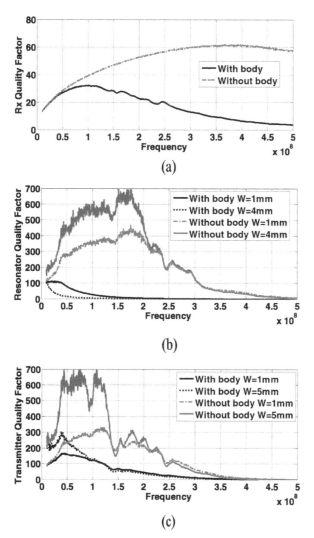

Figure 6.11 The measured quality factor of (a) 1-mm 4-turns square Rx coil, (b) resonator and (c) Tx coil as a function of frequency.

Because of small size of the 1-mm Rx coil, some of the parasitic components imposed by measurement instruments and interconnects have values comparable with the target parameters, such as the coil and mutual inductances. Therefore, measurements should be done carefully by cancelling or attenuating these parasitic effects. One of the common techniques is de-embedding [16], which was adopted by implementing an identical interconnect from the Rx coil to the SMA connector with the distal end, where L_4 is to be located in the actual measurement, shorted. The short and the open circuits de-embedding technique is used to remove the parallel and series parasitic components from the coil and SMA connector while measuring the PTE and PDL. Figure 6.12 shows the effects of interconnect parasitics on the magnitude of the link's transmission coefficient (S_{21}). The specifications of the presented three-coil link is measured and presented in Table 6.1 after de-embedding.

To achieve the maximum PTE in the presence of tissue layers, as Figure 6.5 algorithm suggests, we segmented the Tx coil and resonator by 3 and 2, respectively, resulting in the prototype shown in Figure 6.13. We used a $6 \times 7 \times 8$ cm^3 cube cut out of a fresh lamb head as an *in vitro* model for the human brain and its surrounding layers. The sheep brain was refrigerated for no more than 24 hours and kept for 1–2 hours outside in order to reach the room temperature before experimentation. The total thickness of skin, fat, and bone was 8 mm. Figure 6.13 insets show the Rx coil close up view, complete resonator within the cranial space, and measured results comparing the PTE (from S_{21}) of a conventional 2-coil inductive link and the presented

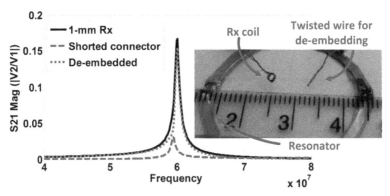

Figure 6.12 De-embedding technique was utilized reduce the effects of EM-field on interconnects from a 1-mm Rx coil to the VNA for PTE measurements and characterization of the 3-coil wireless link.

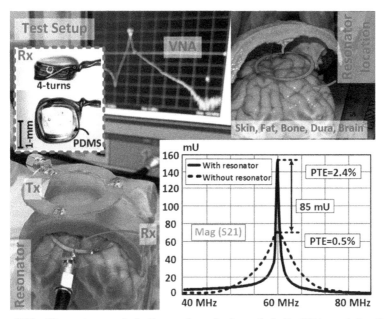

Figure 6.13 The test setup including a sheep brain and skull, VNA, and 3-coil link. Insets clockwise: Rx coil close up view, resonator placement within the cranial space, S_{21} measurements with/without resonator.

3-coil inductive link. The Rx coil Q-factor is low because of its small size and being loaded by the implant electronics. Therefore, it is wideband and relatively insensitive to detuning by the surrounding tissue. Moreover, any undesired shifted in the Rx resonant frequency can be compensated in closed-loop methods, such as automatic resonance tuning (ART) [43]. The resonator, which is unloaded, on the other hand, has a high Q and large surface area. Thus it is more prone to detuning following implantation. Since the resonator is passive, and should be hermetically sealed before implantation, one way to tune it is to carefully estimate the parasitic capacitance, C_{3p}, and the resulting shift in resonance frequency via HFSS simulation and *in vitro* measurements. Then use a lower capacitance, C_{3air}, to tune the resonator in air at a higher frequency for $C_{3tissue} = C_{3air} + C_{3p}$ to resonate at the desired value. In our prototype setup in Figure 6.14, there was 2.6% (1.6 MHz) shift in the resonance frequency of the resonator before and after placement in the tissue, as shown in Figure 6.14, which was pre-tuned for $f = 60$ MHz in the tissue. Placement and removal of the skull in this setup resulted in an additional 0.17% frequency shift.

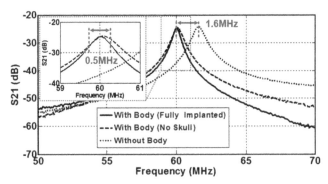

Figure 6.14   Transmission coefficient (S_{21}) of the prototype three-coil link before and after implantation showing the resulting shift in resonance frequency as a result of the parasitic effects of the surrounding tissue.

6.4.2 Wireless Link Measurement Methods

Characteristics of an inductive link vary by the source (input) and load (output) resistors. Although measuring the PTE and PDL using NA and SA is simple, straightforward, and accurate, the measured results are valid only for the ports' impedances at 50 Ω. We present a method that allows using NA and SA for measurement with any desired source and load resistances that match the actual operating conditions. For this purpose, we (i) add resistors in series or in parallel with the NA ports, (ii) measure PTE by NA and PDL by SA (over 50 Ω), and (iii) calculate the actual PTE and PDL. This method provides more flexibility in design and optimization of inductive links by finding the optimal load resistor and model from measurement of the link.

In this method, we add resistances in series or in parallel to the 50 Ω resistances of the measurement instrument ports for the conditions that the actual source and load impedances are higher or lower than 50 Ω, respectively. The presented technique can be used to measure the PDL of an inductive link with any desired load resistor, as well. For PDL measurement, we use a spectrum analyzer (SA), which is matched to 50 Ω. By adding resistors in series or in parallel to the port and measuring the received power over the 50 Ω portion of that circuit, we can calculate the actual amount of delivered power.

To accurately measure the PTE of an inductive link at actual source and load resistances, we propose to use S_{21} of the link while using the circuit configuration presented in Figure 6.15 and associated equations. Each LC-tank is tuned at the carrier frequency, and the entire wireless link whether two- or multi-coil forms a two-port network including any configuration of

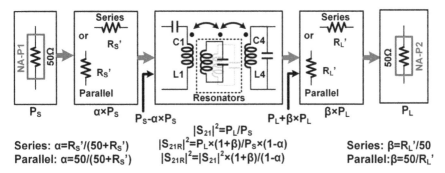

Series: α=R$_S$'/(50+R$_S$')
Parallel: α=50/(50+R$_S$')

$|S_{21}|^2$=P$_L$/P$_S$
$|S_{21R}|^2$=P$_L$×(1+β)/P$_S$×(1-α)
$|S_{21R}|^2$=$|S_{21}|^2$×(1+β)/(1-α)

Series: β=R$_L$'/50
Parallel:β=50/R$_L$'

Figure 6.15 Theory behind the presented PTE and PDL measurement.

resonators between the source (L_1) and load (L_4) coils. L_1 is driven by Port-1 of a Network Analyzer (NA). The default impedance of NA and SA ports is 50 Ω, which can be larger or smaller than the nominal resistance, depending on the applications. For output load of larger than 50 Ω, an additional resistor, R'_L, must be added in series with Port-2 of the NA to ensure $R_L = R'_L + 50$ Ω. On the other hand, if actual R_L is smaller than 50 Ω, an extra resistor must be connected in parallel with Port-2 of the NA, ensuring that the equivalent impedance at the load is $R_L = R'_L \| 50$ Ω. Similarly, for the source impedance, an additional resistor, R'_S, must be added in series or parallel with Port-1 of the NA to ensure either $R_S = R'_S + 50$ Ω or $R_S = R'_S \| 50$ Ω, respectively.

To measure the transmission coefficient under actual load, S_{21R}, we add the extra resistors and then measure S_{21} by NA, which measures the transmission coefficient across the 50 Ω matched load. Finally, we calculate S_{21R} (transmission coefficient of the link with the actual load) and substitute it in PTE general equation,

$$PTE = |S_{21R}|^2 \times 100 \ (\%) . \tag{6.8}$$

Therefore, for series load condition,

$$|S_{21R}|^2 = \frac{V_2'^2}{V_1^2} \times \left(\frac{50}{R_L}\right) = |S_{21}|^2 \times \left(\frac{R_L}{50}\right), \tag{6.9}$$

where V_1 and $V_2'(V_2' = V_2 \times (R'_L + 50)/50)$ are the voltages across NA Port-1 and load coil, respectively, as indicated in the Figure 6.16. V_2 is the voltage across NA's Port-2. For parallel load condition,

$$|S_{21R}|^2 = |S_{21}|^2 \times \left(\frac{R_L}{R'_L}\right) . \tag{6.10}$$

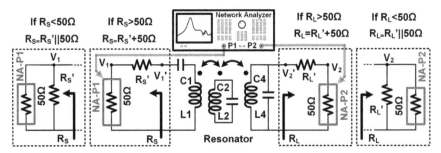

Figure 6.16   Measurement setup configuration including NA and the additional resistors for all the possible source and load conditions.

The Equations (6.9) and (6.10) (as well as Equations (6.11) and (6.12)) are valid only if we measure S_{21} using the configuration in Figure 6.15, where S_{21} is measured over 50 Ω part of the actual load.

The output impendence of the power amplifiers is not necessarily 50 Ω, and 50 Ω is not necessarily the optimal source impedance to achieve the maximum PTE. Therefore, to measure the S_{21} under actual source resistance, S_{21R}, we measure S_{21} by NA, while a proper amount of resistance is added to Port-1, depending on the applications requirement. For series source condition,

$$|S_{21R}|^2 = |S_{21}|^2 \times \left(\frac{R_\mathrm{S}}{50} \right), \tag{6.11}$$

while for parallel source condition,

$$|S_{21R}|^2 = |S_{21}|^2 \times \left(\frac{R_\mathrm{S}}{R'_\mathrm{S}} \right). \tag{6.12}$$

As Equations (6.9–6.12) are summarized in Figure 6.15, any combination of these equations can be used to satisfy application's source and load conditions and measure actual PTE accurately.

To measure PDL with the actual load resistance, we use SA, which has a 50 Ω matched port. Therefore, we use series or parallel R'_L with the 50 Ω load of the SA's port for larger or smaller actual loads, respectively, and calculate the PDL from,

$$\mathrm{PDL} = P_\mathrm{SA} \times (1 + \beta), \tag{6.13}$$

where the P_SA is the power delivered to the 50 Ω load of the SA port and the β equals $R'_\mathrm{L}/50$ or $50/R'_\mathrm{L}$ for series or parallel conditions, respectively (Figure 6.19).

Figure 6.16 presents the proposed measurement setup configuration in detail including NA, the inductive link, and the resistors in parallel or series for both source and load ends. As shown in this figure, to measure the performance of the link for: (i) $R_S < 50 \ \Omega$, R'_S is in parallel with 50 Ω; (ii) $R_S > 50 \ \Omega$, R'_S is in series with 50 Ω; (iii) $R_L < 50 \ \Omega$, R'_L is in parallel with 50 Ω; and (iv) $R_L > 50 \ \Omega$, R'_L is in series with 50 Ω, while the 50 Ω corresponds to the impedance of the NA's ports' 1 and 2. The source and load resistors of an inductive link are determined by specs of power amplifier on the Tx side and the powered application on the Rx side, respectively, which directly influence the design of the link. The presented method can also be used for optimizing an inductive link, during simulation step. Since the HFSS software (ANSYS, Canonsburg, PA, USA) provides scattering parameters, we can use the same configuration in design and calculate the PTE from this software under realistic source and load conditions. The HFSS results also provide a mean to verify the accuracy of the presented method since in this software we can define the desired impedance for each port and directly obtain S_{21R}, without using extra resistors in Equations (6.9–6.12).

Figures 6.17(a,b) show the HFSS model and prototype of a three-coil inductive link, designed for powering mm-sized implants, respectively. The carrier frequency was chosen 60 MHz for this design and the coil geometries are optimized for operating within head tissue layers. In both simulation and measurement, we have swept the load and source resistors from 10 to 1000 Ω and from 10 to 500 Ω, respectively.

6.4.3 Measurement Results

Figure 6.18(a) presents the simulated and measured PTE as a function of R_L. The inset shows the measured S_{21} and calculated S_{21R}, using Equation (6.9),

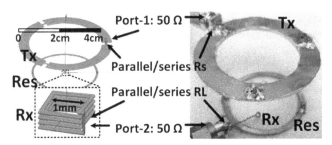

Figure 6.17 (a) The HFSS model and (b) implemented prototype of a three-coil link to verify the presented PTE and PDL measurement method.

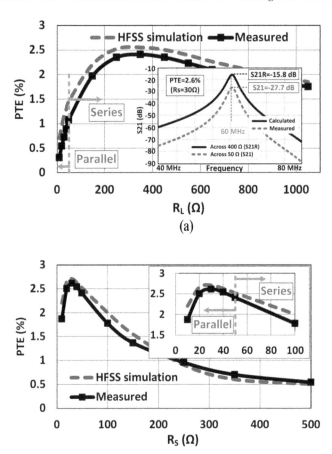

Figure 6.18   Simulated and measured PTE (a) as a function of R_L including S_{21} and S_{21R} vs. frequency in the inset, and (b) as a function of R_S.

as a function of frequency. To generated this curve, we have changed R'_L using discrete resistors from 12 Ω to 1 kΩ. Figure 6.18(b) shows the simulated and measured PTEs as a function of R_S, which indicated that $R_S = 30\ \Omega$ is the optimal source resonance for this design. To measure the PDL at the desired load condition, we use SA to measure the delivered power to its 50 Ω port, in a measurement setup similar Figure 6.16, which is presented in Figure 6.19 (inset). Figure 6.19 also shows the measured PDL using SA as a function of frequency and the actual PDL over R_L, which is calculated from Equation (6.13).

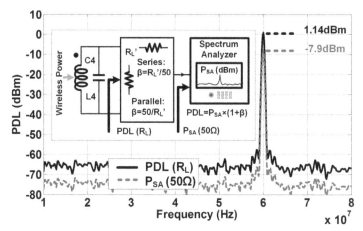

Figure 6.19 Measurement test setup using SA for measuring PDL, and the measured PDL as a function of frequency.

Unlike conventional two-coil inductive link, the measured PTE of the presented three-coil inductive link represents the worst case scenario when the Rx coil is located at the center of the resonator. Nonetheless, it still provides 4.8 times higher PTE than its two-coil counterpart under perfect alignment condition. This is a key advantage of the three-coil link, enabling high PTE across a large area within the high-Q resonator, as shown in Figure 6.4(b). We verified the uniformity and robustness of PTE in the desired coverage area by sweeping the location of the Rx coil along the x- and y-axes across the resonator surface.

We measured S_{21} for both two-coil and three-coil inductive links under similar conditions in Figure 6.16 setup, after de-embedding, and the results are presented in Figure 6.20(a). The three-coil link and segmentation have significantly improved the minimum level and uniformity of the PTE. It can be seen that PTE of the three-coil inductive link increases as the Rx coil moves away from the center of the resonator and closer to its inner perimeter. Therefore, all distributed FFIs receive enough power as long as they are encompassed by the resonator. Moreover, unlike the traditional two-coil inductive links, accurate alignment between the Rx coil(s) and the Tx and resonator coils is not necessary. Even though it is out of the scope of this article, it is possible to have multiple resonators under a larger Tx coil to interface with separate populations of FFIs and brain neural networks. Since the size of the Rx coil is much smaller than the resonator, it does not noticeably influence the magnetic field inside and around the resonator.

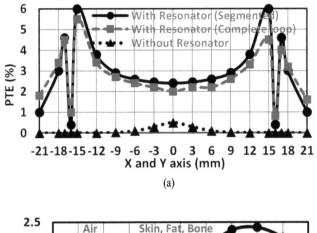

(a)

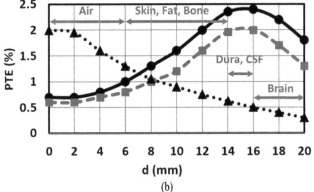

(b)

Figure 6.20   (a) Measured uniformity of the PTE when sweeping the Rx coil across the resonator area, indicating the significant improvement in the PTE in powering the FFIs on the brain at 16 mm depth using the proposed three-coil link, compared to the case without a resonator, and (b) PTE measurement while sweeping the Rx coil depth along z-axis at the center of the Tx and resonator coils.

Instead, it can be considered as an electromagnetic probe that measures the magnetic field relayed by the resonator. According to the Biot–Savart law, the magnetic field intensity drops in inverse proportion to distance from a straight wire, which is the case for Rx coil and resonator when $d_{\mathrm{Rx}} \ll d_{\mathrm{Res}}$. For the area enclosed by the resonator coil, this result in a U-shaped PTE curve, as shown in Figure 6.20(a).

Figure 6.20(b) compares the PTE of the 2-coil and 3-coil inductive links with and without segmentation vs. the distance between a perfectly-aligned

Rx coil and the Tx coil along z-axis, when the resonator for the three-coil inductive link is fixed at $d = 16$ mm. To conduct this test for the range of d from 16 to 20 mm, the surface of the brain was cut at depths of 2 and 4 mm to place the Rx coil below the surface. Although 14 mm $\leq d \leq 16$ mm is the target location for Rx, we also cut the bone and skin to be able to sweep the Rx location across the skull and characterize the PTE as a function of d. As expected, the maximum PTEs for the two-coil and three-coil inductive links along the z-axis are achieved at $d = 0$ mm and $d = 16$ mm, respectively. It can be seen that the PTE in two-coil link is highly sensitive to Tx–Rx separation, while the three-coil link is somewhat flat around 16 mm Tx-resonator distance, for which it has been optimized. Another important observation from this experiment is that in the three-coil inductive link, the PTE drops less than 10% within the designated range of 14–18 mm, indicating that the presented resonance-based three-coil link is not only insensitive to horizontal FFI misalignments within the resonator loop but also quite forgiving in Tx coil distance variations along d. Therefore, as shown in Figure 6.4(b), a large 0.4 cm $\times$ 7 cm^2 = 2.8 cc volume of brain cortical space is considered valid space for placing the FFIs with an average PTE of 3% in this particular prototype.

6.5 Practical Aspects of the FFIs

From practical point of view, it is important to consider the fact that the brain surface is not smooth, and filled with gyri and sulci [49]. Therefore, in our design specifications, the inductive link should also be able to power up the FFIs within $\pm 30°$ angular misalignments. Figure 6.21 provides the simulation and measurement results of PTE and PDL as a function of the Rx coil rotation, which are measured six times for each angle for more accuracy, because of difficulty in adjusting the angular misalignment of Rx in the neural tissue. Error bars show the maximum and minimum values of the measured PDL, while the measured PTE data in this figure represents the average. We can conclude that the presented three-coil link provides PTE $> 1\%$ under angular misalignments of $<\pm45°$. The performance of the presented inductive link is summarized and benchmarked in Table 6.2.

The resonator and the FFIs are surrounded by tissue layers in the cranial space, and all power losses by these components covert to heat and elevate the local temperature. The power that the segmented resonator parasitic resistance dissipates was estimated by measuring the voltage across the resonator, $V_{Res} = 5.9$ V, and $P_{diss} = (V_{Res}/\omega L_{2s})^2 \times R_{2s}/2 = 18.25$ mW for each segment,

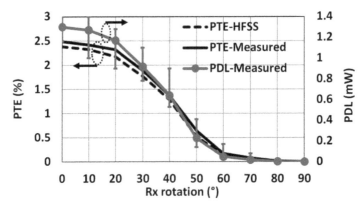

Figure 6.21   HFSS simulation and measurement results of PTE and PDL as a function of Rx rotation, while the Rx is located at the center of the Tx and resonator coils at a depth of 16 mm. Error bars show maximum and minimum of six consecutive measurements.

Table 6.2   Benchmarking wireless power transmission to small implants

Ref.	Link Type	d_{Rx} (mm)	d (mm)	PDL (μW)	PTE (%)	Freq. (MHz)	Core
[14][S]	US[a]	0.1	2	500	7	10	Piezo
[16][M]	2-coil	1	12	224	0.56	200	Air
[44][M]	2-coil	1.15	21.2	7000	0.7	103.7	Ferrite
[45][M]	2-coil	1.6	3	5600	0.18	1500	AP[b]
[46][M]	2-coil	2.2	10	800	0.8	160	Silicon
[47][M]	US	2	80	–	0.04	10	Piezo
[48][M]	2-coil	1	16	76.5	0.68	250	MD[c]
This work[M]	3-coil	1	16	1300	2.4	60	Air

[a]Ultrasound (US), [b]Acrylic Potting (AP), [c]Magnetodielectric (MD).
S: Simulation, M: Measurement.

while delivering 1.3 mW to the Rx. L_{2s} and R_{2s} are the inductance and the resistance of each segment, respectively, equal to the half of the complete loop resonator. While the power loss will raise the temperature in surrounding tissues, blood perfusion takes the heat away and cools down the implantable devices. The power dissipation through the Utah array [50], with size of 55.46 mm^2, increased the temperature linearly with a slope of 0.029°C/mW. Therefore, for a 36.5 mW of power dissipation from both segments of the resonator, over the 100 mm^2 copper surface of the resonator, the estimated temperature rise would be less than 0.3°C, which is well within the 1°C limit. The outer surface of the FFI consisting of a 1 × 1 mm^2 chip and the Rx coil wrapped around it will be 7 mm^2. The temperature rise for maximum delivered power of 1.3 mW would be 0.3°C, and even less (0.23°C) for the target PDL of 1 mW. Adding more FFIs does not change this temperature rise significantly since the outer surface area is also increased linearly with the number of implants.

Considering a uniform distribution of Rx devices in the area enclosed by the resonator, we calculated the number of the Rx devices based on the center-to-center spacing between FFIs. The result is presented in Figures 6.22(a,b) defines the Rx coil's center-to-center spacing, the way it was used in Figure 6.22(a) simulations.

Figure 6.23(a) presents HFSS simulation and measured results of the coupling between a pair of Rx coils, which dimensions are given in Table 6.1, as well as the PTE when the Rx coils are located at the center of the resonator, as a function of the Rx coils' center-to-center spacing. The effects of the surrounding tissue is also considered at 60 MHz. Figure 6.23(b) presents the PDL simulation results for an Rx coil, located at the center of the resonator, as a function of the number of FFIs within the resonator, showing the effect of multiple Rx coils inside the resonator on the PDL and PTE. This figure also presents the overall PTE of all Rx devices as function of the number of FFIs. According to this simulation, the drop in PDL limits the number of FFIs in the system only when the density of devices is very high. The temperature rise in the Rx devices is not dominant and does not limit the number of the Rx devices in the system.

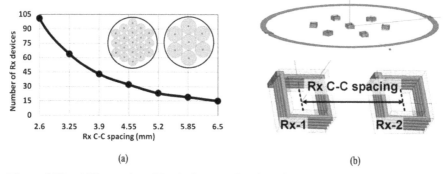

(a)   (b)

Figure 6.22 a) The number of Rx devices as a function of center-to-center spacing between devices that are uniformly distributed inside the area enclosed by the resonator. Red dots: Rx devices, Yellow: The area surrounding each Rx device cleared of other devices. b) HFSS model for seven Rx devices within the resonator.

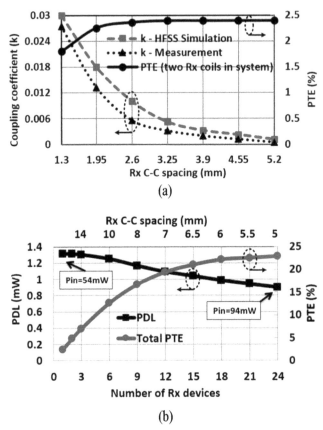

(a)

(b)

Figure 6.23   (a) Simulation and measurement of coupling coefficient between two adjacent Rx coils vs. their center-to-center spacing, and the impact on PTE when they are located at the center of the resonator. Rx coil dimensions are given in Table 6.1. The effects of surrounding tissue is also considered at 60 MHz. (b) PDL of an Rx coil, when it is located at the center of the resonator, and the overall PTE of all Rx coils as a function of the number of FFIs within the resonator and the Rx coils' center-to-center spacing.

6.6 Conclusion

In this chapter, we presented an inductive WPT structure to power up small mm-sized FFIs located within a 3D space inside a high-Q resonator laid on the surface of the brain. The presented three-coil link successfully powered 1 mm-Rx coils within a 2.8-cc volume with an average PTE of 3% at the 60 MHz optimal carrier frequency without surpassing the SAR limits. We have presented HFSS simulation results in support of the design and optimization procedure and segmentation (when beneficial) of the Tx and resonator coils, depending on the input parameters and designated application. We implemented a proof-of-concept prototype to measure and fully characterize the performance of the presented three-coil link in sheep brain in vitro model. Measurement results proved that the presented three-coil link is quite robust against horizontal and angular misalignments of the FFIs as long as they stay with the high-Q resonator. Moreover, the additional heat generated by the resonator coil is within the safe limits.

References

[1] Simeral, J. D., Kim, S. P., Black, M. J., Donoghue, J. P., and Hochberg, L. R. (2011). Neural control of cursor trajectory and click by a human with tetraplegia 1000 days after implant of an intracortical microelectrode array. *J. Neural Eng.* 8:025027.

[2] Egan, J., Baker, J., House, P. A., and Greger, B. (2012). Decoding dexterous finger movements in a neural prosthesis model approaching real-world conditions. *IEEE Trans. Neural Syst. Rehabil. Eng.* 20, 836–844.

[3] Slater, K. D., Sinclair, N. C., Nelson, T. S., Blamey, P. J., and Mcdermott, H. J. (2015). NeuroBi: a highly configurable neurostimulator for a retinal prosthesis and other applications. *IEEE J. Transl. Eng. Health Med.* 3, 1–11.

[4] McMullen, D. P., et al. (2014). Demonstration of a semi-autonomous hybrid brain-machine interface using human intracranial EEG, eye tracking, and computer vision to control a robotic upper limb prosthetic. *IEEE Trans. Neural Sys. Rehabil. Eng.* 22, 784–796.

[5] Wodlinger, B., Downey, J. E., Tyler-Kabara, E. C., Schwartz, A. B., Boninger, M. L., and Collinger, J. L. (2015). Ten-dimensional anthropomorphic arm control in a human brain-machine interface: difficulties, solutions, and limitations. *J. Neural Eng.* 12:016011.

[6] NIH Brain Initiative (2017). *Brain 2025, A Scientific Vision*. Available at: http://www.braininitiative.nih.gov/2025/BRAIN2025.pdf

[7] Schwarz, M. A., Lebedev, T. L., Hanson, D. F., Dimitrov, G. Lehew, J., Meloy, S., et al. (2014). Chronic, wireless recordings of large-scale brain activity in freely moving rhesus monkeys. *Nat. Methods* 11, 760–766.

[8] Brenna, S., Padovan, F., Neviani, A., Bevilacqua, A., Bonfanti, A., and Lacaita, A. L. (2016). A 64-channel 965-μW neural recording SoC with

UWB wireless transmission in 130-nm CMOS. *IEEE Trans. Circuits Syst. II* 63, 528–532.

[9] Shulyzki, R., Abdelhalim, K., Bagheri, A., Salam, M. T., Florez, C. M., Velazquez, J. L. P., et al. (2015). 320-channel active probe for high-resolution neuromonitoring and responsive neurostimulation. *IEEE Trans. Biomed. Circ. Syst.* 9, 34–49.

[10] Song, Y. K., Borton, D. A., Park, S., Patterson, W. R., Bull, C. W., Laiwalla, F., et al. (2009). Active microelectronic neurosensor arrays for implantable brain communication interfaces. *IEEE Trans. Neural Syst. Rehabil. Eng,* 17, 339–345.

[11] Borna, A., and Najafi, K. (2014). A low power light weight wireless multichannel microsystem for reliable neural recording, *IEEE J. Solid State Circuits* 49, 439–451.

[12] Grill, W., Norman, S., and Bellamkonda, R. (2009). Implanted neural interfaces: biochallenges and engineered solutions. *Annu. Rev. Biomed. Eng.* 11, 1–24.

[13] Karumbaiah, L., Saxena, T., Carlson, D., Patil, K., Patkar, R., Gaupp, E., et al. (2013). Relationship between intracortical electrode design and chronic recording function, *Biomaterials* 34, 8061–8074.

[14] Seo, D., Carmena, J. M., Rabaey, J. M., Alon, E., and Maharbiz, M. M. (2013). Neural dust: an ultrasonic, low power solution for chronic brain machine interfaces. *Neurons Cogn.* arXiv:1307.2196.

[15] Lee, B., Ghovanloo, M., and Ahn, D., (2015). Towards a three-phase time-multiplexed planar power transmission to distributed implants. *IEEE Int. Symp. Circuits Syst.* 4, 1770–1773.

[16] Ahn, D., and Ghovanloo, M. (2016). Optimal design of wireless power transmission links for millimeter-sized biomedical implants. *IEEE Trans. Biomed. Circuits Syst.* 10, 125–137.

[17] Biederman, W., Yeager, D., Narevsky, N., Koralek, A., Carmena, J., Alon, E., et al. (2013). A fully-integrated, miniaturized (0.125 mm^2) 10.5 μW wireless neural sensor. *IEEE J. Solid State Circuits* 48, 960–970.

[18] Vullers, R., Schaijk, R., Visser, H., Penders, J., and Hoof, C. (2010). Energy harvesting for autonomous wireless sensor networks. *IEEE Solid State Circuits Mag.* 2, 29–38.

[19] Ho, J. S., Yeh, A. J., Neofytou, E., Kim, S., Tanabea, Y., Patlollab, B., et al. (2014). Wireless power transfer to deep-tissue microimplants. *Proc. Natl. Acad. Sci. U.S.A.* 111, 7974–7979.

[20] Yeh, A. J., Ho, J. S., Tanabe, Y., Neofytou, E., R. Beygui, E., and Poon, A. S. Y. (2013). Wirelessly powering miniature implants for optogenetic stimulation. *Appl. Phys. Lett.* 103:163701.

[21] Seo, D., Tang, H. Y., Carmena, J. M., Rabaey, J. M., Alon, E., Boser, B. E., et al. (2015). Ultrasonic beamforming system for interrogating multiple implantable sensors. *Proc. IEEE Eng. Med. Biol. Conf.* 2015, 2673–2676.

[22] Kurs, A., Karalis, A., Moffatt, R., Joannopoulos, J. D., Fisher, P., and Soljacic, M. (2007). Wireless power transfer via strongly coupled magnetic resonances. *Sci. Express* 317, 83–86.

[23] Mark, M. (2011). *Powering mm-Size Wireless Implants for Brain-Machine Interfaces*. Ph.D. dissertation, University of California, Berkeley, CA.

[24] Ersen, A., Elkabes, S., Freedman, D. S., and Sahin, M. (2015). Chronic tissue response to untethered microelectrode implants in the rat brain and spinal cord. *J. Neural Eng.* 12, 016019.

[25] Nikola, T. (1901). Method of Utilizing Effects Transmitted Through Natural Media. U.S. Patent 685 954.

[26] Tomar, A., and Gupta, S. (2012). Wireless power Transmission: Applications and Components. *Int. J. Eng. Res. Technol.* 1, 5.

[27] Montgomery, K. L., Yeh, A. J., Ho, J. S., Tsao, V., Iyer, S. M., et al. (2015). Wirelessly powered, fully internal optogenetics for brain, spinal and peripheral circuits in mice. *Nat. Methods* 12, 969–974.

[28] Fu, M., Yin, H., Zhu, X., and Ma, C. (2015). Analysis and Tracking of Optimal Load in Wireless Power Transfer Systems. *IEEE Trans. Power Electron.* 30, 3952–3963.

[29] Kiani, M., Jow, U. M., and Ghovanloo, M. (2011). Design and optimization of a 3-coil inductive link for efficient wireless power transmission. *IEEE Trans. Biomed. Circuits Syst.* 5, 579–591.

[30] Kiani, M., and Ghovanloo, M. (2013). A figure-of-merit for designing high performance inductive power transmission links. *IEEE Trans. Ind. Electron.* 60, 5292–5305.

[31] Witricity Inc. Available at: http://witricity.com/

[32] Jow, U. M., and Ghovanloo, M. (2009). Modeling and optimization of printed spiral coils in air, saline, and muscle tissue environments. *IEEE Trans. Biomed. Circuits Syst.* 3, 339–347.

[33] Jow, U. M., McMenamin, P., Kiani, M., Manns, J. R., and Ghovanloo, M. (2014). EnerCage: a smart experimental arena with scalable architecture for behavioral experiments. *IEEE Trans. Biomed. Eng.* 61, 139–148.

[34] Mirbozorgi, S. A., Jia, Y., Canales, D., and Ghovanloo, M. (2016). A wirelessly-powered homecage with segmented copper foils and closed-loop power control. *IEEE Trans. Biomed. Circuits Syst.* 10, 979–989.

[35] Lee, B., Ahn, D., and Ghovanloo, M. (2016). Three-phase time-multiplexed planar power transmission to distributed implants. *IEEE J. Emerg. Sel. Top. Power Electron.* 4, 263–272.

[36] Alghrairi1, M. K., Sulaiman1, N. B., Sidek, R. M., and Mutashar, S. (2016). Optimization of spiral circular coils for bio-implantable micro-system stimulator at 6.78 MHz ISM band. *ARPN J. Eng. Appl. Sci.* 11, 7046–7054.

[37] Schuylenbergh, K. V., and Puers, R. (2009). *Puers: Inductive Powering Basic Theory and Application to Biomedical Systems, Analog Circuits and Signal Processing*. Dordrecht: Springer.

[38] Available at: http://niremf.ifac.cnr.it/tissprop/htmlclie/htmlclie.php

[39] Finkenzeller, K. (2003). *RFID Handbook: Fundamentals and Applications in Contactless Smart Cards and Identification*, 2nd Edn. West Sussex: Wiley.

[40] Qing, X., Goh, C. K., and Chen, Z. N. (2010). A Broadband UHF near-field RFID antenna. *IEEE Trans. Antennas Propag.* 58, 3829–3838.

[41] Furse, C., Christensen, D. A., Durney, C. H. (2009). *Basic Introduction to Bioelectromagnetics*, 2nd Edn. New York, NY: CRC Press.

[42] IEEE (2006). *IEEE Standard for the Safety Levels with Respect to Human Exposure to Radiofrequency Electromagnetic Fields, 3 KHz to 300 GHz, IEEE Standard C95.1*. Piscataway, NJ: IEEE.

[43] Lee, B., Kiani, M., and Ghovanloo, M. (2016). A triple-loop inductive power transmission system for biomedical applications. *IEEE Trans. Biomed. Circuits Syst.* 10, 138–148.

[44] Theilmann, P. T. (2012). *Wireless Power Transfer for Scaled Electronic Biomedical Implants*. Ph.D. thesis, University of California, San Diego, CA.

[45] Montgomery, K. L., Yeh, A. J., Ho, J. S., Tsao, V., Iyer, S. M., Grosenick, L., et al. (2015). Wirelessly powered, fully internal optogenetics for brain, spinal and peripheral circuits in mice. *Nat. Methods* 12, 969–974.

[46] Zargham, M., and Gulak, P. G. (2015). Fully integrated on-chip coil in 0.13 µm CMOS for wireless power transfer through biological media. *IEEE Trans. Biomed. Circuits Syst.* 9259–271.

[47] Meng, M., Ibrahim, A., and Kiani, M. (2015). Design considerations for ultrasonic power transmission to millimeter-sized implantable microelectronic devices. *IEEE Biomed. Circuits Syst. Conf.* 1–4.

[48] Moradi, E., Björninen, T., Sydänheimo, L., Ukkonen, L., and Rabaey, J. M. (2013). Antenna design for implanted tags in wireless brain machine interface system. *Proc. IEEE Antennas Propag. Soc. Intl. Symp.* 2083–2084.

[49] Yang, F., and Krugge, F. (2008). Automatic segmentation of human brain sulci. *Med. Image Anal.* 12, 442–451.

[50] Kim, S., Tathireddy, P., Normann, R. A., and Solzbacher, F. (2007). Thermal impact of an active 3-D microelectrode array implanted in the brain. *IEEE Trans. Neural Syst. Rehabil. Eng.* 15, 493–501.

7

Next Generation Neural Interface Electronics

Ian Williams, Lieuwe Leene and Timothy G. Constandinou

Imperial College London, UK

7.1 Introduction

The enormous potential of neural interfaces – as prosthetics, therapeutics and as scientific tools – is only just a beginning to be tapped. These tools have an incredible breadth of application and have already shown life-changing capability for amputees, quadriplegics, and those suffering from neurological diseases [1, 2]. However, even some of these amazing examples struggle to be translated into clinical practice because of challenges in stability, repeatability, cost, and requirements for skilled technicians or restrictions on motion due to the use of benchtop equipment. It is clear that the interfaces of today fall far short of the performance that researchers and clinicians dream of delivering.

Transitioning from today's devices to systems capable of delivering even a fraction of the potential of neural interfaces, will require more than just the incremental technological change associated with Moore's law (and its equivalents in for example wireless or battery technology). It will require visions to drive new research – identifying new goals and alternative paths for achieving these goals. Here, we present an example vision of the next generation of neural interfaces and look at how well-existing research trends are addressing it, the limits of these incremental trends, and finally at interesting alternative approaches that could leapfrog known limits.

7.1.1 Today's Neural Interfaces

Artificial neural interfaces arguably started with Galvani's electrical stimulation experiments, and electrical neuromodulation (both stimulation and blocking) remains the primary implantable clinical application. Deep brain stimulation (DBS) devices for treating neurological disorders linked to particular brain states (e.g., Parkinsons, depression, or epilepsy disease) and devices for chronic pain are the most common devices, and typically consist of a low number of electrodes connected to a relatively simple biphasic pulse generator where the frequency and amplitude of stimulation is set by a human operator. Cochlear implants, in contrast, are more complex with more electrodes (typically <25) and with both the frequency of the pulse trains, and the electrode on which they are produced controlled electronically, according to the sounds picked up by the device's microphone.

Electrical recording is only just on the verge of breaking through in clinical applications – DBS devices for epilepsy or Parkinson's are starting to include neural recording to identify the changes in brain state that presage or indicate the onset of symptoms – enabling the stimulation to be modified and thereby improving efficacy, power efficiency, and reducing unwanted side effects [3–5].

7.1.2 The Neural Interface of the Near Future

Our vision of the next generation of implantable neural interfaces requires implantable devices capable of:

- Stably recording for years from thousands of neurons; interpreting these signals in realtime to identify behavioural and motor states or sensory inputs; and initiating appropriate neuromodulation in a neurally and therapeutically relevant timescale (in the order of milliseconds) or wireless transmitting identified state change events to external systems.
- Stably initiating naturalistic pulse trains in thousands of target neurons (within the mixed population of a major nerve) with negligible activation of non-target neurons and with automatic calibration of stimulation parameters or invariant safe and effective whole nerve stimulation.
- Long term stimulation or block of a single neuron or target neural network portion in the brain.

Key aspects of achieving this stable performance include the electrodes and the implant packaging, however, this chapter focuses on the electronics aspects. In depth reviews and recent developments of recent electrode development can be found in [29, 45, 71, 90, 114] and of packaging can be found in [35, 96, 98, 104, 108]. There are also basic architectural challenges involved in scaling-up to hundreds or thousands of channels – for instance floating electrodes (i.e., with wires between the electrodes and electronics) already suffer from high interconnect failure rates and these challenges will increase exponentially as the channel count rises and connector and wire sizes decrease. Two options for high channel count devices are clear in the literature, the first is to simply move away from floating electrodes – instead using microfabrication techniques to bond electrodes directly to the neural interface IC – the second is to use a distributed architecture comprising many small devices wirelessly linked together [99].

The driving factors that should enable the brain–machine interfaces BMI community to meet these objectives are outlined in Figure 7.1. This illustrates that there are primarily two domains that drive innovation for improved BMI systems. An essential aspect of this progress is inherently due to basic neuroscience. By increasing our understanding of the nervous system and by finding new applications and sensing modalities, these electronic systems can become more effective at coupling to the human body. This includes a broad field of research-oriented objectives exploring different sensory mechanisms and neurophysiology. In association to these efforts technology translation of electronic systems has also led to very rapid developments in the past decade. There are numerous examples where this is pushing the capabilities

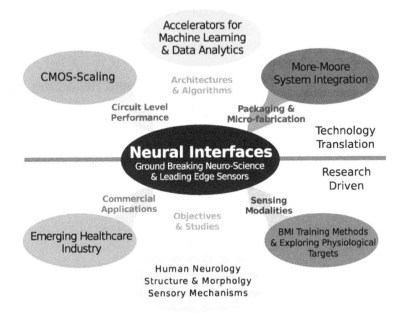

Figure 7.1 Primary factors influencing development of leading edge neural interfaces.

of current integrated systems by utilizing state of the art signal processing techniques and CMOS micro fabrication. A crucial aspect is that the enabling electronics plays an equally significant role as the scientific breakthroughs from biomedical research. For this reason, we highlight three factors for improvement: circuit level performance, architectures and algorithms, and microfabrication.

7.2 Sensing Neural Activity

Since its inception in the 1920s, electrophysiology has remained the primary method of monitoring neural activity. The design of CMOS-based electrophysiology systems has developed into a well-established art with many leading edge systems still resembling foundational work by Harrison et al. [40]. The systems for cortical implants and micro-electrode arrays (MEAs) in particular, strive to deliver exceptional monitoring of neural populations by densely integrating a very large number of instrumentation circuits on chip. The basis for this aim is the strong evidence that increasing the number of simultaneous recordings for predicting complex motor control directly improves the decoding accuracy of the BMI [100]. This section will discuss an interesting sensor challenge that is emerging as the recording capacity is perpetually increasing. In extension, we shall detail a number of emerging developments that have enabled exciting opportunities for BMIs.

7.2.1 Power Density Target for High Channel Count Recording

The trend towards increased recording capacity emphasises the importance for careful sensor design that maintains adequate thermal dissipation. Particularly, when current systems extensively perform processing on chip. It remains to be the case that electrode configurations like the Utah array [14, 39] are distributed over a volume of cortex while sensing take place on a two dimensional (2D) array of instrumentation circuits. Clearly, fabrication technologies could provide 3D integration of sensor circuits. However, this may actually exhibit worse thermal capacity than 2D integration due to reduced surface area per recording unit. The key requirement is then that integrated devices should realize very high 2D density in order to probe the different layers and columns of cortical tissue.

In order to reveal the challenge of managing thermal capacity experienced by state of the art recording systems, we have surveyed a number of sensor systems published over the past 5 years. This is summarized in Figure 7.2

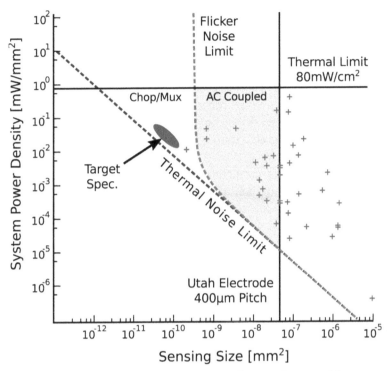

Figure 7.2 System power density with respect to sensing area for state of the art recording systems.

which illustrates how well high-density recording is achieved with respect to the thermal budget of 80 mW/cm^2 [120]. In particular, the system power density is considered with respect to the sensor density. The power density is evaluated in terms of the total system power dissipation and the fabricated silicon die size which typically dissipates its heat into the surrounding tissue. On the other hand, the sensor size is taken with respect to the size of integrated electrode pitch or the size of the corresponding instrumentation circuits in order to emulate the achievable electrode pitch.

Probably, the most interesting aspect of this depiction is the constraint imposed by the thermal noise limit. This results from the minimum current dissipation from each sensing circuit in order to achieve 5 μV_{rms} input referred noise given a 5-kHz bandwidth. A strict amount of power needs to be dissipated irrespective of the instrumentation size resulting in a power density inversely proportional to sensor area. The closer a system is to this limit the better the NEF it will exhibit. Unless the supply voltage, noise requirement, or bandwidth is readjusted the system power density cannot fall below this limit. As a consequence, managing the thermal budget is an imperative design consideration for future systems that target extreme miniaturization. This is usually required in order to minimize device impact on biological tissues. For such a scenario, reducing the acquisition bandwidth may become a predominant reason for using Local Field Potential (LFP) recordings over spiking activity. This could be the only way to allow several thousand channels to be integrated onto a sub-millimetre size CMOS device without thermal concerns.

The flicker noise sources for active readout transistors also plays an important role that can impair such achievements. If we assume transistors are equal in size to that of the sensing area, then it should be expected that below a certain size the flicker components inhibit signal detection. Conventional readout topologies cannot achieve very high densities due to flicker noise if chopping or electrode multiplexing techniques are not utilized. However, such structures guarantee DC blocking behaviour which is desirable for safety regulations. It is well known that chopper instrumentation can achieve substantially smaller configurations but degrades instrumentation input impedance [20]. The system in Johnson et al. [44] shows how the use of chopping can achieve 50 μm electrode pitch for exceptional spacial resolution with 768 active recording channels. This particular configuration performs quantization at the sensor interface using a ramp ADC and a open loop amplifying structure to minimize the number of analogue components in the system. A similar structure is employed in [49] with the addition of digital feedback in order to introduce high-pass filtering behaviour. This key modification allows robust *in vivo* recording that is not impaired by electrode drift or large fluctuation in local field potentials that are typically not present *in vitro*.

Another means to accommodate higher sensing densities is achieved by actively multiplexing the electrode array and only selecting the most informative electrode locations for the decoding task. This allows the work in Ballini et al. [6] to interface over 26 k electrodes while simultaneously recording from 1024 units for the *in vitro* study of cell cultures. By embedding SRAM memory cells within the electrode switch matrix this system presents a highly versatile MEA platform that can perform extensive signal conditioning. The main challenge is that the system complexity results in a 76 mm^2 die size which is difficult to translate towards an implantable solution. For this reason the work in Lopez et al. [62] focuses explicitly using a silicon probe as substrate to allow 52 simultaneous recording channels from 455 electrodes.

This approach seems to be one of the most promising for future BMI systems as the electrodes can be fabricated together with the electronics using well established micro-fabrication [97].

In addition to these high-density recording systems, a number of promising circuit techniques have been proposed that may allow considerable improvements in performance. For instance, the application of bulk switching in Han et al. [38] is capable of reducing flicker noise without sacrificing the DC-blocking characteristics. On the other hand [72] suggests removing the capacitive coupling structure altogether by extensively utilizing digital feedback which could achieve very high-recording densities if electrode multiplexing is also incorporated. Overall this trend towards higher recording densities is expected to remain a sustained effort where both fabrication and circuit techniques will play a key role for miniaturization.

7.2.2 Emerging Technologies

The current focus of innovation is predominantly directed at achieving exceptional instrumentation performance in combination with diversifying system functionality. This includes a number of capabilities such as; wireless telemetry, closed loop stimulation, and signal characterization to name a few. It is becoming apparent that increasing the capacity current BMIs requires us to find more efficient means to perform signal extraction as well as finding more effective interfacing strategies. For this reason we will highlight a few of the promising technologies that will hopefully make substantially more effective systems in the future.

7.2.2.1 Advanced CMOS technologies

A number of recent BMI publications show a growing interest for using advanced CMOS technologies in order to accommodate more digital processing capabilities on chip. This is particularly relevant for closed loop neuromodulation that needs more sophisticated diagnostics to perform therapeutic feedback. To some extent the high channel count recording systems also necessitate extensive processing for various types of signal compression. As a result there are a number of opportunities associated with using nanometre CMOS processes. For example the 0.25V neural processor in Liu and Rabaey [60] is able to perform feature extraction on quantized recordings with exceptional power efficiency in part due to the 65 nm technology. Combining this with existing sensor interfaces that operate at very low supply voltages [37, 73] may result in an order of magnitude improvement in power dissipation when compared to previously proposed systems.

It is important to point out that there are other challenges that prevent conventional instrumentation structures to deliver precise sensing at these technology nodes. Particularly, when attempting to achieve a compact configuration with good linearity at these reduced supply voltages [112]. It should not be surprising that the use of oscillators have been particularly successful to leverage the digital design style [46]. Our group recently demonstrated that oscillator based structures, while high digital in nature, can allow exceptional performance for filtering time domain signals [59]. In fact, the concept of encoding signals in the time domain has been proposed in a number of recent works [27, 69, 123]. This is motivated by asynchronous processing capabilities for sparse neural activity that could drastically reduce power dissipation [34]. However, this is still an ongoing effort where the current realizations show exceptional dynamic range but have not yet been able to demonstrate the same noise efficiency conventional methods. This is partly because such benefits can only be realized when both instrumentation and signal processing is performed using the time domain signal modality. Some further argument can be made that the reduced parasitics and smaller geometries from this trend make chopper circuits substantially more viable. This is because the input capacitance can be reduced if the closed loop gain is maintained relatively large. Such a reduction should translate towards boosting the sensor's input impedance to hundreds of mega ohm. Again this works in favour of VCO topologies because they inherently exhibit excessive open-loop gain.

7.2.2.2 Chemical sensing

In addition to the electrical activity that can be recorded from neural populations, chemical markers such as dopamine concentrations can play a key role in utilizing neurochemistry to refine the diagnostic fidelity of BMIs. The neurochemostat is a recent innovation that allows a means to perform closed-loop regulation of endogenous neurotransmitters [11]. Because a number of neuropathologies relate explicitly to deficiencies in specific neurotransmitters this development can enable new therapeutic strategies that probe different pathways using rich neurochemistry. While chemical sensors face other challenges associated to the electrode composition and interface that can can degrade sensitivity in chronic settings [74]. However, there are a number of promising means to prevent such degradation such as anti-fouling coatings [65]. Moreover, a number of existing sensor platforms already allow simultaneous sensing of electrical and chemical activity with hundreds of different channels to study neurodegenerative diseases [33].

7.2.2.3 Optical sensing

The use of optics in implantable sensors is possibly one of the newer themes in brain machine interfaces as a result of the enabling success in optogenetics. Such sensors may provide essential clinical tools to precisely guide neuro-surgery as well as new high resolution imaging tools for brain structures [22]. That said there are also efforts to perform label-free imaging sensors that use the polarization of reflected light which does not require optogenetic transfection [122]. While there is a great amount of functional flexibility, implantable optics face a demanding challenge in relation to developing compact devices. This is exemplified by the implantable prototype in Paralikar et al. [82] which uses a coupled fibre to deliver optical stimulation from a battery powered system. This is primarily an outstanding challenge for recording activity from large volumes [43] that has the potential to be integrated along side more conventional recording circuits [44].

7.3 Signal Processing

There are many levels of possible resolution for recording neural signals. At one extreme is the activity of single neurons (e.g., using a patch clamp electrode) or at the other extreme it is possible to study broad activity of neuron populations (e.g., of the brain with EEG). Today's implanted devices generally focus on single neuron activity or compound activity of a relatively small sub-population of neurons using electrodes microns to a few millimetres away from the neurons. The signal content picked up from these electrodes is illustrated in Figure 7.3.

7.3.1 Single Neuron Processing

Singling out the neural firing pattern of a single neuron chronically is challenging. Intracellular microelectrodes can be used in the short term, but the neuron under test typically dies rapidly. An alternative approach is to record signals extracellularly and use signal processing to separate the contribution of different neurons in the vicinity [25, 66, 92]. This is possible if the action potentials are significantly above the general noise floor of the recordings and because variations in the type, orientation, and proximity of the neurons around the electrode each influence the shape and amplitude of action potentials (spikes) recorded from that neuron.

The process of separating spikes from different neurons is called spike sorting and typically consists of three stages (Figure 7.4): (i) filtering the

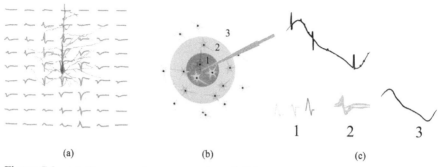

(a)   (b)   (c)

Figure 7.3   (a) Diagrammatic representation of different spike shapes recorded around a neuron depending on electrode Paralikar neuron distance and relative orientation. (b) The recorded neural signal is a composite of high SNR signals from nearby neurons (1), low-SNR multi-unit activity from slightly more distant neurons (2) and a local field potential (LFP) resulting from the sum activity of many distant neurons (3). (c) The recorded neural signal broken down into the 3 components described in (b).

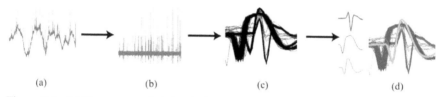

(a)   (b)   (c)   (d)

Figure 7.4   (a) The raw recorded signal, (b) after filtering, (c) the spikes all peak aligned (d) the spikes after clustering.

raw signal to remove the LFP and any artefacts (e.g., 50 or 60Hz power line interference); (ii) identifying spikes from amongst the remaining noise – a process called spike detection; and (iii) using algorithms (or manually) grouping spike shapes with similar characteristics to identify how many neurons are differentiably contributing to the recorded signal (called clustering), these groupings can then give a series of deinterleaved spike trains from the original spike train.

Spike detection is typically achieved using simple amplitude threshold crossing and with this kind of approach the challenge is in setting the threshold appropriately. There can be very substantial differences in noise and neural signal amplitudes, so methods for automatically calculating this threshold have been proposed [51, 86], however, an alternative approach is to use operators that are less sensitive to absolute amplitude (such as non-linear energy operators [53]) however, these are typically more complex to implement.

After detection the clustering process can begin. This is typically a computationally intensive process of feature extraction and statistical analysis. Implementations on embedded hardware and ASICs have been proposed [47, 48, 83, 95] and achieving best performance within the power, memory and computational power constraints of these systems remains a key challenge. Careful end to end system design is critical [7, 8, 24, 78], however, the real time and tight constraints of these systems necessarily limit the performance and flexibility achievable.

Once clustering has been completed and each spike has been assigned to a particular neuron (cluster), then deinterleaved spike trains for each of the neurons found can be extracted. Interpreting these spike trains (either in real-time for closed loop applications or offline for analytical applications) typically involves binning the spikes into timeslots, determining the spike rate and setting firing rate thresholds that indicate atypical behaviour that correlates with underlying patient/animal activity.

In many neuroscience applications it is strongly desired to be able to record single unit neural activity for long periods of time in freely behaving animals, and it is also desirable to be able to trigger neural stimulation or some external action in rapid response to patterns of single unit activity (within a few milliseconds). The key challenges with achieving these aims, are the high data rates associated with raw data recording (which drive power consumption for wireless transmission and data logging) and the computational complexity of clustering and deinterleaving the data to identify single unit activity. In our work on the NGNI v1 platform we have developed a small, low power headstage (Figures 7.5 and 7.6) capable of both streaming out 32 channels of raw recorded data and of performing real time deinterleaving of neural data on those 32 channels [63, 119]. The key operating principle

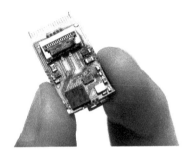

Figure 7.5 The NGNI v1 headstage.

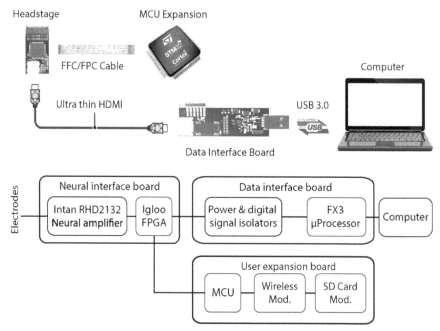

Figure 7.6 The NGNI v1 hardware showing the headstage with its connections either to a microcontroller or via a USB link to a computer.

of this device is that it uses a two-stage process (Figure 7.7), in stage 1 the device is tethered to a computer and raw data is recorded and processed offline – this leverages the high performance unsupervised spike train clustering and deinterleaving capability of a proprietary WaveClus implementation – to create templates (a characteristic waveform for neuron's spike); then in stage 2 we use those templates to perform spike sorting in real time on the headstage by using a computationally simple template matching (or feature recognition) process and output a deinterleaved spike stream for either wireless transmission, SD card logging or triggering some form of activity. How well this two phase approach works depends in particular on the stability of the waveform shape, both over time as electrodes become encapsulated, but also during more complex behaviour e.g. bursting. Performance is shown in Table 7.1 and indicates that we can decode the activity of 128 neurons with sub-millisecond latency and around 30 mW power consumption. Initial in-vivo trials have been successfully performed in non-human primates and long term recording trials will start soon targeting first 1 day and then 1 week of operation.

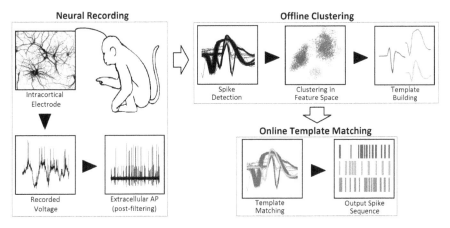

Figure 7.7 The two-phase spike sorting process. In the first phase the neural signal recording is processed offline to generate templates, in the second phase the neural data is passed to an FPGA and the templates are used to perform real time spike sorting.

Table 7.1 NGNI v1 performance measures

Channels	32	Templates/channel	4
Input range (LSB–full range)	0.4 μV– 26 mV	Input Impedance	13 MΩ @ 1 kHz
Sampling rate	15 kSample/s	Reference Input Impedance	0.5 MΩ @ 1 kHz
Spike detection method	Single Threshold	Amplifier Input Referred Noise	2.4 μVrms
Classification method	Template matching	Low-frequency 3dB Cut-off	0.02–500 Hz
Alignment method	Peak	High-frequency 3dB Cut-off	100–20 kHz
Template length	16 samples (1ms)	Max spike sorting rate	100 kSpikes/s
Distance measure	Σ (absolute difference)	Max spike sorting latency	300 μs

7.3.2 Population Level Processing

Action potentials are believed to be the language of individual neurons, however, they are difficult to record chronically due to the instability of current electrodes, the high raw data rates and computational complexity of deinterleaving the spike train.

If action potentials are the conversations between individual neurons, then there is an emerging belief that for many applications sufficient information can be gleaned simply from the tone of the local conversation. By measuring this tone or LFP – believed to be a summation of synaptic activity of a nearby population of neurons – it is hoped that a simpler more stable interface can be achieved [13, 26, 36].

The advantages of using LFPs stem from three aspects of the signal and the signal source: (i) the frequency band the signal covers (typically in the order of 1–100 Hz) which is over an order of magnitude less than spikes and so leads to correspondingly lower data rates and substantial potential power savings; (ii) the minimal impact of gliosis and encapsulation of the electrode on the LFP signal; and (iii) for some applications the key information content is believed to simply be in the phase-amplitude coupling of several distinct frequency bands which can be readily extracted (e.g., using the Fast Fourier Transform or heterodyning approaches) [3, 4, 15, 21, 26, 36]. However, LFPs are not without their own unique challenges: the shift in frequency causes an increase in $1/f$ noise as well as a reduction in temporal resolution, and for many applications the signal is inherently ambiguous and identifying the relationship between LFPs and either the underlying neural sources or the animal's movement/sensation requires substantial computational processing and potentially investigator insight. Also although the frequency bands of interest are broadly agreed there is significant variation in the exact frequency definitions [26].

7.3.3 Signal Processing Summary

The advantages of LFPs are attractive, however, in many current neuroscience applications, single unit activity remains the preferred parameter. Looking to the near future it may well be that with more tool development (to enable an experimenter to rapidly go from raw LFP data to an output that can be used experimentally) that LFPs will become preferred for chronic applications where latencies in the tens to hundreds of milliseconds are acceptable.

7.4 Neuromodulation

Our vision of near-future neurotherapeutics and neuroprosthetics requires hundreds to thousands of electrodes implanted in the Central or Peripheral Nervous System (CNS or PNS) with a high percentage achieving a significant and stable neural effect. In fact current systems in animal models can already demonstrate a level of locomotion recovery following spinal cord transection using closed loop neuromodulation techniques [116].

To date, clinical neural modulation devices have typically utilised electrical stimulators with low numbers of electrodes and areas of activation in the order of 1 mm^3 – several cm^3. The surgeon implanting the devices will typically choose where to insert the device/electrodes using either feedback from an awake patient or by targeting visible anatomical features, thereby giving an accuracy of stimulation targeting commensurate with the activation area. In the research domain stimulators capable of thousands of channels of independent activation exist as do micro-electrode arrays with hundreds to thousands of channels, however, key challenges remain such as selectivity, specificity, stability, and uncertainty.

To illustrate this consider the implantation of a micro-electrode array. Apart from knowledge about the area of innervation of a nerve/area of cortex, the insertion is done largely blind – the neural tissue is essentially a black box to which tens to thousands of connections have now been formed. Further the relationship between electrodes and neurons is not one-one, but rather many-many (both for stimulation and recording).

The experimenter must typically therefore gain information about the neurons a particular electrode has interfaced with by either: correlating recorded signals to patient actions and stimulations or manipulations of the patients body; or by sending a series of stimulations into the electrode and observing the physical reaction or noting any sensations the patient describes. The process of stimulating is, however, not necessarily a straight forward one, many possible parameters can be adjusted such as stimulation amplitude, stimulation duration, inter-phase duration, stimulation frequency or waveform. This is further complicated because not all afferent neurons give rise to liminal sensations, and the many-many relationship of the electrode neuron interface means that different settings may give rise to different subsets of neurons being activated, so potentially large parameter sweeps are required on each electrode. In a many channel (N channels), electrode this can take a significant amount of patient and experimenter time. In a bipolar stimulator where stimulation can be between any two electrodes (modifying the activation area) this parameter space is multiplied by $^{N}C_2$ and in a multipolar stimulator up to $^{n}C_M$ combinations are possible (where M is the number of channels involved in each stimulation) although in practice simpler pre-determined combinations would likely be used.

Assuming this calibration phase is successful, then the experimenter will have a mapping between electrode stimulation settings and physical effects. These physical effects will be dominated by muscles/receptors innervated by large fibre diameter neurons due to electrical stimulation preferentially stimulating these neurons (dominant factors are axon size and axon proximity to the electrode). Therefore, the usefulness of the mapping will in part be determined by whether this recruitment order suits the desired application. However, for many applications (e.g., smooth recruitment of muscle or stimulation of certain sensations) there is a need to selectively or preferentially stimulate small fibres. This issue can in part be addressed by using techniques like waveform modulation [31] and speed selective blocking [54, 87], but so far with limited effectiveness.

Assuming that a useful mapping can be achieved the experimenter then has a potentially short time for which this mapping will be valid. In particular when micro-electrode arrays are used, they often introduce a break in the blood-nerve barrier and may cause endoneural accumulation of fluid that leads to neural damage. The body's tissue rejection reaction also causes fibrous build up around the electrodes, insulating them and pushing neurons away. In addition, the mechanical incompatibility (in particular the density and stiffness of rigid electrodes) causes relative motion between the electrodes and the tissue and in particular a boring effect at the electrode tip [10, 85]. Flexible electrodes using polymer substrates have been shown to reduce these effects, and in theory biocompatible or bioactive coatings could also greatly reduce tissue rejection, however, these techniques are still in their infancy [55, 117].

To summarize there are a number of key issues with using a high channel count neural stimulator such as: the biocompatibility of the electrodes, the large and unstable search space for stimulation settings, the lack of ability to target specific neuron types, the limited control of current flow in the tissue (combined with a lack of knowledge about the physical distribution of the target neurons in the tissue). The situation is not necessarily so bleak in all areas of the body, e.g., in specialized receptor areas such as the cochlear or retina there is much greater homogeneity of receptor type and a relatively clear functional and anatomical distribution of these receptors, however, each area of neural tissue brings its own unique challenges.

Another point to note is that in the clinical field, blocking is of major interest, however, it appears to receive significantly less research interest than stimulation. Here, we make a point of talking of neuromodulation to capture both blocking and stimulation.

Our vision of a next generation neuromodulation device requires hundreds of stable channels with neuron-type specificity and high selectivity as well as low-power blocking and stimulation. The rest of this chapter provides a very brief summary of neuromodulation device design, identifies existing trends in these devices (primarily from the research domain) and limits to these trends, as well as looking at emerging technologies and approaches which could circumvent these limits and move us towards next generation performance.

7.4.1 Existing Approaches: Issues and Challenges

Neural stimulation ICs can be grouped by their stimulation front-ends as either current controlled, voltage controlled or charge controlled – as depicted in Figure 7.8 and described in Chapter 4.

7.4.1.1 Current controlled stimulators

Numerous designs of biphasic current pulse generators have been proposed in the research literature. Recent designs have started to share many common features including:

- A stable and temperature insensitive current reference. Relatively little work has been published using high-accuracy current references, potentially reflecting the low importance placed on the absolute current values used for stimulation (except from a safety perspective).
- A 4–8 bit current DAC depending on how wide a range of stimulation currents desired. The linearity of these is generally only focused on when using asymmetric biphasic pulses (for charge balancing purposes).
- A current mirror output stage with high output impedance, current scaling (to reduce power waste) and a low minimum voltage drop.
- One or more charge balancing mechanisms (as discussed in Section 7.4.2.3).

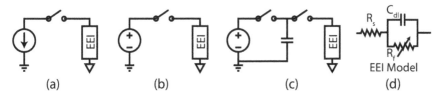

Figure 7.8   Diagrammatic representations of output stages for neural stimulators.

7.4.1.2 Voltage controlled stimulators

Voltage controlled stimulators are simpler than current controlled devices – typically consisting mainly of a DAC and buffer. They are more common in clinical devices than in the research literature, likely due to their simplicity as a medical device, proven effectiveness and low-power operation.

7.4.1.3 Charge controlled stimulators

Charge controlled stimulators are of growing popularity in the research domain, but have not yet made a move into clinical devices and are potentially difficult to scale due to the need for external capacitors as well as charging time and unknown discharging time.

7.4.2 Trends and Limits

A number of trends are apparent in a review of the literature on neural stimulation ICs. In particular there have been readily apparent increases in the number of stimulation channels and conversely reductions in the per channel size and power consumption. In addition, there are a number of non-numeric trends, such as changes in charge balancing approaches, and increasingly System on Chip (SoC) levels of integration.

7.4.2.1 Channel count and size trends

Over the last few decades there has been a clear and rapid rise (roughly tenfold per decade) in the number of channels on a single implantable neural stimulation IC as indicated in Figure 7.9. A single data point for *in vitro* IC is included indicating the current potential capability for integration and with the stimulation field shaping capabilities of bipolar and multipolar stimulation this number can be boosted further by the creation of virtual channels. However, if we restrict the analysis to implantable devices and real channels then the number of channels per IC has plateaued over the last few years at around 4000, and if we further eliminate devices targeting the retina then it becomes clear that this levelling off began a decade earlier at around 100 channels.

The likely underlying cause of this levelling off is the diminishing returns from high channel counts due to: (i) the paucity of stable and biocompatible electrodes with more than 100 channels; (ii) charge spreading effects which reduces the effective channel count when using small pitch electrodes (e.g., using bondpads on the IC as electrodes); (iii) the durability of interconnects for arrays with high electrode numbers; and (iv) the generally black

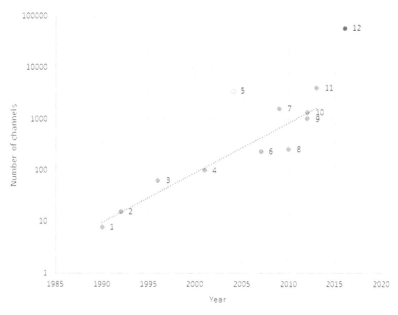

Figure 7.9   Trend in channel number per IC stimulator. Green and red devices not included for trend line calculation – green dot indicates an ineffective device, red is an *in vitro* device included to indicate limits of currently achievable integration. Devices: 1–12 can be found in [16, 17, 50, 56, 75, 79, 81, 93, 94, 105, 107, 113].

box nature of the nervous system (although the retina and cochlea present relatively well understood and spatially structured interface).

As the channel count has increased there has naturally been a corresponding decrease in the per-channel size. In contrast to neural recording, a minimal pixel size can be relatively simply achieved – i.e., using current steering or row/column decoders it can involve just a few transistor switches per electrode. The limits on this size reduction depend on the stimulation voltages being used (potentially requiring the use of stacked low voltage or large high voltage transistors) and the stimulation current (large width transistors may be required to reduce the voltage drop across them). Currently, a limiting factor on this size reduction is the minimum size and pitch of connections off-chip (e.g., bondpads or Through Silicon Vias).

7.4.2.2 Power consumption trends

Current controlled stimulators remain the most common design found in the research literature, however, in clinical applications (such as DBS) voltage controlled devices dominate with proven efficacy and power efficiency trumping the potential safety benefits of charge control. Power waste in a current controlled neural stimulator is largely a result of the voltage drop across the current controlling transistors in the output stage (Figure 7.10 (b)). As such there has been a move to adaptive power supplies to try and minimise these voltages (see Figure 7.10 (c, d)). The ideal supply would track the potential difference across the electrodes with a constant minimal amount of headroom to keep the transistors in saturation, however, in practice designs have tended to use discrete voltage levels whch are easier to generate without

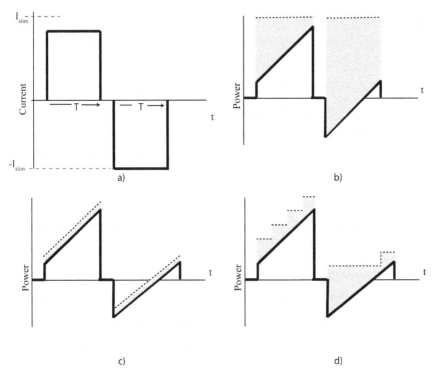

Figure 7.10 (a) A biphasic current pulse. (b) Dotted lines indicate typical stimulator power consumption, solid line indicates power necessary to drive the current through the electrodes, the grey area indicates power waste. (c) An ideal current controlled stimulator would continuously modulate the voltage across the current controlling transistors to minimise power waste (although some minimal headroom is likely necessary to keep the transistors in saturation). (d) Typical solutions in the literature have made use of discrete positive voltage levels leading to power consumption [118].

external components. It should be noted that due to the overheads associated with the voltage conversion, energy savings generated by proposed designs are heavily dependent on: front-end topology, stimulation current and stimulation waveform. Further it should be noted that although papers typically quote power savings in excess of 50% this may be relative to a baseline system efficiency of only 10% (i.e., taking efficiency up to 20%).

7.4.2.3 Charge balancing

Biphasic stimulation pulses which leave the electrode at a electrochemically neutral potential difference to the surrounding solution (minimizing long-term Faradaic reactions) has been a mainstay of neural stimulator design for decades. Early techniques focused on passive techniques such as DC-blocking capacitors and electrode shorting, while later devices in the research field looked at precision equalising of the charge delivered in the two phases (Figure 7.11). However, studies have cast doubt on all of these approaches [61, 102, 111]. A currently favoured approach involves monitoring electrode voltages and delivering or removing excess charge to achieve target voltages [61,103]. Ultimately however, the use of stimulation voltages outside the water window will always lead to unwanted electrochemical reactions with implications for both tissue and electrode health.

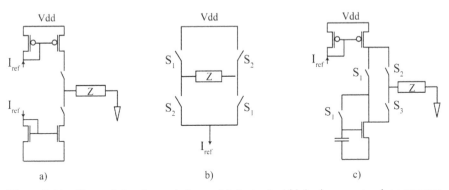

Figure 7.11 Charge balancing techniques. (a) A standard biphasic current pulse generator – mismatch between the reference currents and mirrors leads to charge imbalance. (b) An H-bridge enables the use of a single reference current – greatly enhancing charge balance [19]. (c) A current-copying calibration approach, during calibration just switches S_1 are closed, copying the PMOS sourced current, then during stimulation switches S_1 are open and switches S_2 and S_3 are utilized [101].

7.4.3 System on Chip

The fully integrated realization of neuromodulation ICs has been an outstanding endeavour for a number of research efforts that is rapidly becoming more successful. As demonstrated by Medtronic's pioneering work [58] that is one of the few to realize neuromodulation devices with FDA approval. Similar closed loop systems extensively use LFP signals for seizure prevention have been particularly successful at realizing integrated solutions. For instance the system in Yoo et al. [121] uses extensive segregation of different LFP frequency bands from 0.5 to 30 Hz as features which allows a support vector machine in the digital baseband to to perform classification on EEG activity. While this is not quite as invasive a solution as the Medtronic device this work illustrates that the processing architecture plays a key role in on-chip resource requirements. Generally the level of invasiveness will reflect in the system's latency to detect events in neural activity while exhibiting more challenging instrumentation requirements for EEG systems when compared to ECoG or intracortical recordings. In fact the newer implementation in Bin Altaf and Yoo [2] uses a non-linear basis SVM engine to improve its hardware efficiency resulting in an accuracy improvement from 84 to 95%.

To some extent the integrated form factor allows these SOCs to deliver substantially increased sensing and stimulation capabilities when compared to benchtop alternatives. In fact the overall experimental complexity is considerably less and environmental noise sources are less likely to perturb an integrated system. For instance the system in Abdelhalim et al. [1] uses 64 recording and stimulation channels for cortically implanted electrode arrays to deliver electrical stimulation therapy based on closed loop control. This type of active neuromodulation can improve clinical efficacy in freely moving rats. Earlier results [5] indicated that such an approach can reduce epileptic seizures by 90% in addition to improving chronic viability owning to a reduced implant size and wireless capability. This particular system uses phase synchrony in specific LFP frequency bands to trigger current mode stimulation with minimal feedback latency. A similar wireless system in Rhew et al. [88] also uses LFP energies to deliver stimulation but proposes to take advantage of log-based signal encoding in neural activity to improve energy efficiency. This is just one of many examples where the physiology of neural activity can be used to allow more effective processing methods. The target application here is using deep brain stimulation to treat essential tremor and Parkinson's disease using a programmable PI controller.

An isolated number of works have used neural spiking activity to realize closed loop stimulation control of neural activity on SOC platforms. This is possibly due to the large stimulation artefacts that can disturb the instrumentation front end but also because most target applications are directed at external motor control. The system in Biederman et al. [9] suggests adiabatic charge-recycling may potentially realize better efficiencies. However the recorded spike rates are not directly used to adjust the stimulation pattern without off-chip intervention. On that note the system in Mendez et al. [67] utilizes the spiking rates from classified neurons to infer bladder volume and thereby provide a well informed condition for stimulation. This should indicate that peripheral nerve control is more adequate for this class of devices simply due to the lack of information in local field fluctuations. Instead chemical measures or the associated nerve activity is used for closed loop control.

7.4.4 Alternative Approaches and Modalities

Neuromodulation is fundamentally about either initiating or blocking the propagation of action potentials. Action potential initiation is typically caused by depolarizing part of the neural cell membrane beyond a critical threshold level (typically in the order of $\sim$15 mV). Once depolarized beyond this level a positive feedback loop is (caused by the opening of sodium channels) which turns the small depolarization into a full reverse polarization. This depolarizes the surrounding membrane and a chain reaction ensues that is observed as an action potential propagating along the neuron. Conversely blocking is usually achieved by either hyperpolarizing the cell membrane, or by altering the states of sodium, potassium or calcium channels and pumps to prevent action potential initiation.

The flow of current into a portion of the cell to initiate an action potential, can be modulated in a number of ways such as: changing the transmembrane voltage, or by altering the membrane capacitance, or by affecting the flow of ions through the membrane. Clinical neural stimulators utilise direct electrical stimulation to change the transmembrane voltage, however, this presents significant challenges at the electrode/electrolyte boundary and poor selectivity or stimulation targeting capability (due to the unknown and changing tissue structure, charge spreading and preferential recruitment of larger axons). However, alternative stimulation methods exist that could circumvent these limits and help deliver our vision. In this section we will be briefly covering 3 alternative modalities: optogenetics, acoustic and thermal. Other modalities such as chemical or magnetic stimulation have been demonstrated but are outside the implantable and electronic focus of this book, a review can be found in [41, 64].

7.4.4.1 Optogenetics

Optogenetics in some ways presents the clearest path to targeted neural stimulation, with selection possible by opsin expression (which can be targeted spatially and to a particular neuron type), by light amplitude, and by wavelength. The incredible sub-cellular spatial resolution of this technique enables action potentials to be generated or blocked in a single neuron at a time and the temporal resolution likewise exceeds practical requirements. However, the safety of genetically modifying human cells remains a major challenge. There have recently been several human trials approved which should be invaluable in demonstrating the safety and efficacy of the technique in humans. These trials focus on the CNS as questions remain about sufficiency of expression and stability of expression in the non-immune privileged PNS. To date, implanted device capability has been very limited and remains hampered by major thermal challenges from light intensity requirements, LED inefficiencies, and system designs which offer little control over the light intensity delivered to the tissue – optogenetic implants in general lag significantly behind the incredible performance possible with benchtop lasers, fibre optics and sophisticated light targeting setups.

7.4.4.2 Acoustic/Mechanical

Neuromodulation by the On–Off Keying (OOK) of ultrasound has been demonstrated as a possible technique for retinal and brain cells [68, 76, 77, 110]. The mechanisms for stimulation and inhibition are poorly understood, but is thought to be due to mechanical deformation of neural cell membranes and ion channels (with cavitation and thermal effects begin investigated). These cavitation and thermal effects are also a cause for concern as the establishment of standing waves (or other focusing effects) could cause damaging levels of energy to be delivered to parts of the tissue.

The potential performance of ultrasound for neural stimulation is still being established. Research by Menz et al. [68] indicated an *in vitro* spatial resolution of $\sim100\,\mu$m utilizing high-frequency transducers (48 MHz), however, spatial resoution *in vivo* is likely to be significantly degraded (in the order of a few millimeteres) [77]. Temporal resolution is harder to pinpoint although latencies of tens of milliseconds have been reported which is relatively poor in comparison to electrical or optogenetic stimulation. However, a key advantage of acoustic stimulation is the capability to utilize it without implantation and potentially through bone (although spatial resolution will be substantially degraded).

At this point in time it seems likely that ultrasound will remain a tool for non-invasive neuromodulation as well as for transcutaneous power and data transfer to implanted devices.

7.4.4.3 Thermal

Stimulation and blocking of action potentials has been demonstrated using spatially localised heating of neural tissue. The difference in effect (stimulation or blocking) is believed to derive from two conflicting mechanisms – ion channel dynamics and membrane capacitances – which are affected by changes in temperature [23, 84].

Rapid local heating causes capacitive effects to dominate and can initiate action potentials, this is because the rate of change in membrane capacitance (dC_m/dt) and hence the stimulus magnitude is proportional to the speed of temperature change (dT/dt), and can simply be expressed as: $dC_m/dt = (dC_m/dT)\,(dT/dt)$. In contrast, with slower heating the changes in ionic conductances (especially Na^+ and K^+ activation/deactivation dynamics) dominate and blocking is achieved [23, 70].

Thermal modulation has primarily been achieved with Infra-Red (IR) lasers and in acute studies. This approach enables stimulation with greater spatial and temporal resolution than direct electrical stimulation and also eliminates the stimulation artefact (ideal for electrical neural recordings). However, achieving similar effect with an implantable device remains an open challenge and thermal tissue damage from chronic use remains a major concern. Future *in vitro* and *in vivo* testing is likely necessary before functional IR implants can be considered [89, 115]. An exciting development in thermal modulation is the introduction of nanoparticles that can be preferentially bound to certain neurons or neuron types – greatly improving the ability to target delivery of thermal excitation [18, 42].

7.4.5 Neuromodulation Summary

Outside of specialised sensory regions such as the retina, it is becoming clear that simply scaling the number of channels of electrical stimulation brings diminishing returns due to the limited targeting possible during electrode insertion and the generally black box nature of the nervous system. There also remain significant challenges with respect to long term biocompatibility and stability and it is likely that novel electrode materials and design are necessary to deliver substantial improvements in these areas, (although advances in stimulator ASIC design have already delivered improvements and novel approaches in managing Faradaic currents at the electrode-electrolyte interface).

Optogenetics and thermal stimulation in particular offer the tantalising opportunity to sidestep many of the problems associated with electrical stimulation, however, they both are relatively new fields with limited efficacy data in humans and bring their own problems for implantation – both for instance struggle with thermal damage – and for optogenetics there are all the problems associated with genetic modification and the innate immune response. More information on these alternative approaches and others can be found in Woong Lee et al. [57] and luan et al. [64].

Achieving our vision (of many channels of highly targeted and stable stimulation) is likely to require significant advancements in fields outside of electronics design and some form of additional targeting (such as precision electrode placement or changes in target neuron excitability, e.g., through genetic or nanoparticle means). An interesting development is the increasing use of multiple stimulation modalities in a single implant, e.g., combining optogenetics with IR nanoparticle stimulation or electrical stimulation to leverage the advantages of the different techniques and mitigate the weaknesses [18, 64].

7.5 Discussion

The market for neural interfaces is already growing rapidly and we believe that delivering our vision of highly targeted neuromodulation controlled by detailed knowledge of neural network state would both greatly enhance existing applications and open up exciting new opportunities.

The trends for neural interface implant development are clear – channel counts are growing and devices are becoming smaller, lower power and more integrated. There are also a continuous stream of neuroscience discoveries which enhance our understanding of the nervous system and how best to interface to it; as well as major projects (such as the Human Brain Project and Human Connectome Project) which could transform our understanding of the brain and open up entirely new avenues of research. However, major challenges remain, in particular: (i) biocompatibilty of the implant and the electrode-electrolyte interface; (ii) the black box nature of the system at the time of implantation; and (iii) the limited specificity and selectivity of todays neuromodulation implants.

In this chapter we have discussed new approaches and methods that are coming online and how they could feed into our vision, but there are also new ways of working (e.g., the growing open hardware movement epitomised by the founding of the Open-Ephys group and the HardwareX journal or collaborations between formerly unrelated tech giants such as Google and GSK) and there is always the potential that developments in a seemingly unrelated field could reshape the neural interfaces field (such as chronic percutaneous links being developed in osseointegration research [80] or the growing of artificial nerves in tissue engineering [12, 28, 32]).

Our vision places a particular emphasis on scaling up the number of target neurons interfaced with and ensuring that the interface is stable. It seems clear that achieving this will require developments outside of the electronics field, particularly in electrode design and materials or alternative interfacing modalities (especially, optogenetics for neuromodulation and chemical for sensing) and enhanced methods of targeting particular neuronal subpopulations. The next generation of neural interfaces will not be achieved with simple Moore's law gains, but rather through close interdisciplinary working and experimentation.

References

[1] Abdelhalim, K., Jafari, H. M., Kokarovtseva, L., Velazquez, J. L. P., and Genov. R. (2013). 64-channel uwb wireless neural vector analyzer soc with a closed-loop phase synchrony-triggered neurostimulator. *IEEE J. Solid State Circ.* 48, 2494–2510.

[2] Bin Altaf, M. A., and Yoo, J. (2016). A 1.83 µj/classification, 8-channel, patient-specific epileptic seizure classification soc using a non-linear support vector machine. *IEEE Trans. Biomed. Circ. Syst.* 10, 49–60.

[3] Aru, J., Aru, J., Priesemann, V., Wibral, M., Lana, L., Pipa, G., Singer, W., and Vicente, R. (2015). Untangling cross-frequency coupling in neuroscience. *Curr. Opin. Neurobiol.* 31, 51–61.

[4] Al-Thaddeus, A., Santa, W., Carlson, D., Jensen, R., Stanslaski, S., Helfenstine, A., and Denison, T. (2008). A 5 w/channel spectral analysis IC for chronic bidirectional brain–machine interfaces. *IEEE J. Solid State Circ.* 43, 3006–3024.

[5] Bagheri, A., Gabran, S. R. I., Salam, M. T., Perez Velazquez, J. L., Mansour, R. R., Salama, M. M. A., and Genov, R. (2013). Massively-parallel neuromonitoring and neurostim-ulation rodent headset with nanotextured flexible microelectrodes. *IEEE Trans. Biomed. Circ. Syst.* 7, 601–609.

[6] Ballini, M., Mller, J., Livi, P., Chen, Y., Frey, U., Stettler, A., et al. (2014). A 1024-channel cmos microelectrode array with 26,400 electrodes for recording and stimulation of electrogenic cells *in vitro*. *IEEE J. Solid State Circ.* 49, 2705–2719.

[7] Barsakcioglu, D. Y., Eftekhar, A., and Constandinou, T. G. (2013). Design optimisation of front-end neural interfaces for spike sorting systems. in *Proceedings of the 2013 IEEE International Symposium on Circuits and Systems (ISCAS2013)*, Rome, 2501–2504.

[8] Barsakcioglu, D. Y., Liu, Y., Bhunjun, P., Navajas, J., Eftekhar, A., Jackson, A., et al. (2014). An analogue next generation neural interface electronics front-end model for developing neural spike sorting systems. *IEEE Trans. Biomed. Circ. Syst.* 8, 216–227.

[9] Biederman, W., Yeager, D. J., Narevsky, N., Leverett, J., Neely, R., Carmena, J. M., et al. (2015). A 4.78 mm 2 fully-integrated neuromodulation soc combining 64 acquisition channels with digital compression and simultaneous dual stimulation. *IEEE J. Solid State Circ.* 50, 1038–1047.

[10] Biran, R., Martin, D. C., and Tresco, P. A. (2005). Neuronal cell loss accompanies the brain tissue response to chronically implanted silicon microelectrode arrays. *Exp. Neurol.* 195, 115–126.

[11] Bozorgzadeh, B., Schuweiler, D. R., Bobak, M. J., Garris, P. A., and Mohseni, P. (2016). Neu-rochemostat: A neural interface soc with integrated chemometrics for closed-loop regulation of brain dopamine. *IEEE Trans. Biomed. Circ. Syst.* 10, 654–667.

[12] Bryson, J. B., Machado, C. B., Crossley, M., Stevenson, D., Bros-Facer, V., Burrone, J., et al. (2014). Optical control of muscle function by transplantation of stem cell–derived motor neurons in mice. *Science*, 344, 94–97.

[13] Buzsáki, G., Anastassiou, C. A., and Koch, C. (2012). The origin of extracellular fields and currents? eeg, ecog, lfp and spikes. *Nat. Rev. Neurosci.* 13, 407–420.

[14] Campbell, P. K., Jones, K. E., Huber, R. J., Horch, K. W., and Normann, R. A. (1991). A silicon-based, three-dimensional neural interface: manufacturing processes for an intracortical electrode array. *IEEE Trans. Biomed. Eng.* 38, 758–768.

[15] Canolty, R. T., and Knight, R. T. (2010). The functional role of cross-frequency coupling. *Trends Cogn. Sci.* 14, 506–515.

[16] Chen, K., Yang, Z., Hoang, L., Weiland, J., Humayun, M., and Liu, W. (2010). An integrated 256-channel epiretinal prosthesis. *IEEE J. Solid State Circ.* 45, 1946–1956.

[17] Chow, A. Y., Pollack, J. S., Packo, K. H., and Schuchard, R. A. (2005). The artificial silicon retina microchip for the treatment of retinitis pigmentosa: 2 to 4 1/2 year update. *Invest. Ophthalmol. Vis. Sci.* 46, 2005.

[18] Colombo, E., Feyen, P., Antognazza, M. R., Lanzani, G., and Benfenati, F. (2016). Nanoparticles: a challenging vehicle for neural stimulation. *Front. Neurosci.* 10:2016.

[19] Constandinou, T. G., Georgiou, J., and Toumazou, Christofer. (2008). A partial-current-steering biphasic stimulation driver for vestibular prostheses. *IEEE Trans. Biomed. Eng.* 2, 106–113.

[20] Denison, T., Consoer, K., Santa, W., Avestruz, A. T., Cooley, J., and Kelly, A. (2007). A 2 μw 100 nv/rthz chopper-stabilized instrumentation amplifier for chronic measurement of neural field potentials. *IEEE J. Solid State Circ.* 42, 2934–2945.

[21] Denison, T., Santa, W., Jensen, R., Carlson, D., Molnar, G., and Avestruz, A-T. (2008). "An 8μw heterodyning chopper amplifier for direct extraction of 2μv rms neuronal biomarkers," in *Proceedings of the 2008 IEEE International Solid-State Circuits Conference-Digest of Technical Papers*, (Piscataway, NJ: IEEE), 162–603.

[22] Dombeck, D. A., Khabbaz, A. N., Collman, F., Adelman, T. L., and Tank, D. W. (2007). Imaging large-scale neural activity with cellular resolution in awake, mobile mice. *Neuron* 56, 43–57.

[23] Duke, A. R., Jenkins, M. W., Lu, H., McManus, J. M., Chiel, H. J., and Jansen, E D. (2013). Transient and selective suppression of neural activity with infrared light. *Sci. Rep.* 3:2600.

[24] Eftekhar, A., Sivylla, E. P., and Timothy, G. C. (2010). "Towards a next generation neural interface: Optimizing power, bandwidth and data quality," in *Proceedings of the 2010 Biomedical Circuits and Systems Conference (BioCAS)*, (Piscataway, NJ: IEEE), 122–125.

[25] Einevoll, G. T., Franke, F., Hagen, E., Pouzat, C., and Harris, K. D. (2012). Towards reliable spike-train recordings from thousands of neurons with multielectrodes. *Curr. Opin. Neurobiol.* 22, 11–17.

[26] Einevoll, G. T., Kayser, C., Logothetis, N. K., and Panzeri, S. (2013). Modelling and analysis of local field potentials for studying the function of cortical circuits. *Nat. Rev. Neurosci.* 14, 770–785.

[27] Elia, M., Leene, L. B., and Constandinou, T. G. (2016). "Continuous-time micropower interface for neural recording applications," in *IEEE Proceedings of the International Symposium on Circuits and Systems*, (Piscataway, NJ: IEEE), 534–537.

[28] Faroni, A., Mobasseri, S. A., Kingham, P. J., and Reid, A. J. (2015). Peripheral nerve regeneration: experimental strategies and future perspectives. *Adv. Drug Deliv. Rev*. 82, 160–167.

[29] Fekete, Z. (2015). Recent advances in silicon-based neural microelectrodes and microsystems: a review. *Sens. Actuators B Chem*. 215, 300–315.

[30] Fisher, R. S., and Velasco, A. L. (2014). Electrical brain stimulation for epilepsy. *Nat. Rev. Neurol*. 10, 261–270.

[31] Grill, W. M., and Mortimer, J. T. (1995). Stimulus waveforms for selective neural stimulation. *IEEE Eng. Med. Biol. Magaz*. 14, 375–385.

[32] Gu, X., Ding, F., and Williams, D. F. (2014). Neural tissue engineering options for peripheral nerve regeneration. *Biomaterials* 35, 6143–6156.

[33] Guo, J., Ng, W., Yuan, J., Li, S., and Chan, M. (2016). A 200-channel area-power-efficient chemical and electrical dual-mode acquisition IC for the study of neurodegenerative diseases. *IEEE Trans. Biomed. Circ. Syst*. 10, 567–578.

[34] Guo, N., Huang, Y., Mai, T., Patil, S., Cao, C., Seok, M., Sethumadhavan, S., and Tsi-vidis, Y. (2016). Energy-efficient hybrid analog/digital approximate computation in continuous time. *IEEE J. Solid State Circ*. 51, 1514–1524.

[35] Gwon, T. M., Kim, C., Shin, S., Park, J. H., Kim, J. H., and Kim, S. J. (2016). Liquid crystal polymer (lcp)-based neural prosthetic devices. *Biomed. Eng. Lett*. 6, 148–163.

[36] Hall, T. M., Nazarpour, K., and Jackson, A. (2014). Real-time estimation and biofeedback of single-neuron firing rates using local field potentials. *Nat. Commun*. 5:5462.

[37] Han, D., Zheng, Y., Rajkumar, R., Dawe, G. S., and Je, M., (2013). A 0.45 v 100-channel neural-recording ic with sub-μw/channel consumption in 0.18μm cmos. *IEEE Trans. Biom. Circ. Syst*. 7, 735–746.

[38] Han, M., Kim, B., Chen, Y. A., Lee, H., Park, S. H., Cheong, E., et al. (2015). Bulk switching instrumentation amplifier for a high-impedance source in neural signal recording. *IEEE Trans. Circ. Syst. II* 62, 194–198.

[39] Harrison, R. R., Watkins, P. T., Kier, R. J., Lovejoy, R. O., Black, D. J., Greger, B., et al. (2007). A low-power integrated circuit for a wireless 100-electrode neural recording system. *IEEE J. Solid State Circ.* 42, 123–133.

[40] Harrison, R. R., and Charles, C., (2003). A low-power low-noise cmos amplifier for neural recording applications. *IEEE J. Solid State Circ.* 38, 958–965.

[41] Henry, R., Deckert, M., Guruviah, V., and Schmidt, B. (2015). Review of neuromodulation techniques and technological limitations. *IETE Techn. Rev.* 33, 1–10, 2015.

[42] Huang, H., Delikanli, S., Zeng, H., Ferkey, D. M., and Pralle, A. (2010). Remote control of ion channels and neurons through magnetic-field heating of nanoparticles. *Nat. Nanotechnol.* 5, 602–606.

[43] Ji, N., and Smith, S. L. (2016). Technologies for imaging neural activity in large volumes. *Nat. Neurosci.* 19, 1154–1164.

[44] Johnson, B., Peace, S. T., Wang, A., Cleland, T. A., and Molnar, A. (2013). A 768-channel cmos microelectrode array with angle sensitive pixels for neuronal recording. *IEEE Sens. J.* 13, 3211–3218.

[45] Jorfi, M., Skousen, J. L., Weder, C., and Capadona, J. R. (2015). Progress towards biocompatible intracortical microelectrodes for neural interfacing applications. *J. Neural Eng.* 12:011001.

[46] Karkare, V., Chandrakumar, H., Rozgic, D., and Markovic, D. (2014). "Robust, reconfigurable, and power-efficient biosignal recording systems," in *IEEE Proceedings of the Custom Integrated Circuits Conference*, Austin, TX, 1–8.

[47] Karkare, V., Gibson, S., and Markovic, D. (2011). A 130uw, 64-channel neural spike-sorting dsp chip. *IEEE J. Solid State Circuits* 46, 1214–1222.

[48] Karkare, V., Gibson, S., and Marković, D. (2013). A 75-μw, 16-channel neural spike-sorting processor with unsupervised clustering. *IEEE J. Solid State Circuits* 48, 2230–2238.

[49] Kassiri, H., Bagheri, A., Soltani, N., Abdelhalim, K., Jafari, H. M., Salam, M. T., et al. (2016). Battery-less tri-band-radio neuro-monitor and responsive neurostimulator for diagnostics and treatment of neurological disorders. *IEEE J. Solid State Circuits* 51, 1274–1289.

[50] Kim, C., and Wise, K. D. (1996). A 64-site multishank cmos low-profile neural stimulating probe. *IEEE J. Solid State Circuits* 31, 1230–1238.

[51] Kim, S., and McNames, J. (2007). Automatic spike detection based on adaptive template matching for extracellular neural recordings. *J. Neurosci. Methods* 165, 165–174.

[52] Kingwell, K. (2012). Neural repair and rehabilitation: Neurally controlled robotic arm enables tetraplegic patient to drink coffee of her own volition. *Nat. Rev. Neurol.* 8, 353–353.

[53] Koutsos, E., Paraskevopoulou, S. E., and Constandinou, T. G. (2013). "A 1.5 μw neo-based spike detector with adaptive-threshold for calibration-free multichannel neural interfaces," in Proceedings of the Circuits and Systems (ISCAS), 2013 IEEE International Symposium on, (Kuala Lumpur: IEEE), 1922–1925.

[54] Kuffler, S. W., and Vaughan Williams, E. M. (1953). Small-nerve junctional potentials. the distribution of small motor nerves to frog skeletal muscle, and the membrane characteristics of the fibres they innervate. *J. Physiol.* 121(2):289–317.

[55] Leach, J. B., Achyuta, A. K., and Murthy, S. K. (2010). Bridging the divide between neuroprosthetic design, tissue engineering and neurobiology. *Front. Neuroeng.* 2:18.

[56] Lee, C., and Hsieh, C. (2013). A 0.8-v 4096-pixel cmos sense-and-stimulus imager for retinal prosthesis. *IEEE Trans. Electron Devices* 60, 1162–1168.

[57] Lee, J. W., Kim, D., Yoo, S., Lee, H., Lee, G. H., and Nam, Y. (2015). Emerging neural stimulation technologies for bladder dysfunctions. *Int. Neurourol. J.* 19, 3–11.

[58] Lee, S. B., Lee, H. M., Kiani, M., Jow, U. M., and Ghovanloo, M. (2010). An inductively powered scalable 32-channel wireless neural recording system-on-a-chip for neuroscience applications. *IEEE Trans. Biomed. Circuits Syst.* 4, 360–371.

[59] Leene, L. B., and Constandinou, T. G. (2016). "A 0.45v continuous time-domain filter using asynchronous oscillator structures," in *Proceedings of the IEEE International Conference on Electronics, Circuits and Systems*, Rome: IEEE.

[60] Liu, T. T., and Rabaey, J. M. (2013). A 0.25 v 460 nw asynchronous neural signal processor with inherent leakage suppression. *IEEE J. Solid State Circuits* 48, 897–906.

[61] Lo, Y. K., Hill, R., Chen, K., and Liu, W. (2013). "Precision control of pulse widths for charge balancing in functional electrical stimulation", in *Proceedings of the Sixth International IEEE/EMBS Conference Neural Engineering (NER)*, Rome: IEEE, 1481–1484.

[62] Lopez, C. M., Andrei, A., Mitra, S., Welkenhuysen, M., Eberle, W., Bartic, C., et al. (2014). An implantable 455-active-electrode 52-channel cmos neural probe. *IEEE J. Solid State Circuits* 49, 248–261.

[63] Luan, S., Williams, I., De Carvalho, F., Jackson, A., Quian Quiroga, R., and Constandinou, T. G. (2016). *Next Generation Neural Interfaces for Low-Power Multichannel Spike Sorting*. FENS.

[64] Luan, S., Williams, I., Nikolic, K., and Constandinou, T. G. (2014). Neu-romodulation: present and emerging methods. *Frontiers in Neu-roeng*. 7:27

[65] Machado, R., Soltani, N., Dufour, S., Salam, M. T., Carlen, P. L., Genov, R., and Thompson, M. (2016). Biofouling-resistant impedimetric sensor for array high-resolution extracellular potassium monitoring in the brain. *Biosensors* 6:53.

[66] Martínez, J., and Quian Quiroga, R. (2013). Spike sorting 4. Principles of Neural Coding, page 61,.

[67] Mendez, A., Belghith, A., and Sawan, M. (2014). A dsp for sensing the bladder volume through afferent neural pathways. *IEEE Trans. Biomed. Circuits Syst.* 8, 552–564.

[68] Menz, M. D., Oralkan, O., Khuri-Yakub, P. T., and Baccus, S. A. (2013). Precise neural stimulation in the retina using focused ultrasound. *J. Neurosci.* 33, 4550–60.

[69] Mohan, R., Yan, L., Gielen, G., Hoof, C. V., and Yazicioglu, R. F. (2014). 0.35 v time-domain-based instrumentation amplifier. *Electron. Lett.* 50, 1513–1514.

[70] Mou, Z., Triantis, I. F., Woods, V. M., Toumazou, C., and Nikolic, K. (2012). A. simulation study of the combined thermoelectric extracellular stimulation of the sciatic nerve of the xenopus laevis: the localized transient heat block. *IEEE Trans. Biomed. Eng.* 59, 1758–1769.

[71] Müller, J., Ballini, M., Livi, P., Chen, Y., Shadmani, A., Frey, U., et al. (2013). "Conferring flexibility and reconfigurability to a 26,400 microelectrode cmos array for high throughput neural recordings," in *Proceedings of the 2013 Transducers & Eurosensors XXVII: The 17th International Conference on Solid-State Sensors, Actuators and Microsystems (TRANSDUCERS & EUROSENSORS XXVII)*, Rome: IEEE, 744–747.

[72] Muller, R., Gambini, S., and Rabaey, J. M. (2012). A 0.013 mm^2, 5μw, dc-coupled neural signal acquisition ic with 0.5 v supply. *IEEE J. Solid State Circuits* 47, 232–243.

[73] Muller, R., Le, H. P., Li, W., Ledochowitsch, P., Gambini, S., Bjorni-nen, T., et al. (2015). A minimally invasive 64-channel wireless μecog implant. *IEEE J. Solid State Circuits* 50, 344–359.

[74] Myers, E. B., and Roukes, M. L. (2011). Comparative advantages of mechanical biosensors. *Nat. Nanotechnol.* 6, 1748–3387.

[75] Naganuma, H., Kiyoyama, K., and Tanaka, T. (2012). "A 37× 37 pixels artificial retina chip with edge enhancement function for 3-d stacked fully implantable retinal prosthesis," in *Proceedings of the 2012 IEEE Biomedical Circuits and Systems Conference (BioCAS)*, Rome: IEEE, 212–215.

[76] Naor, O., Hertzberg, Y., Zemel, E., Kimmel, E., and Shoham, S. (2012). Towards multifocal ultrasonic neural stimulation ii: design considerations for an acoustic retinal prosthesis. *J. Neural Eng.* 9:026006.

[77] Naor, O., Krupa, S., and Shoham, S. (2016). Ultrasonic neuromodula-tion. *J. Neural Eng.* 13:031003.

[78] Navajas, J., Barsakcioglu, D. Y., Eftekhar, A., Jackson, A., Constandi-nou, T. G., and Quian Quiroga, R. (2014). Minimum requirements for accurate and efficient real-time on-chip spike sorting. *J. Neurosci. Methods* 230, 51–64.

[79] Noorsal, E., Sooksood, K., Xu, H., Hornig, R., Becker, J., and Ort-manns, M. (2012). A neural stimulator frontend with high-voltage compliance and programmable pulse shape for epiretinal implants. *IEEE J. Solid State Circuits* 47, 244–256.

[80] Ortiz-Catalan, M., Håkansson, B., and Brånemark, R. (2014). An osseointegrated human-machine gateway for long-term sensory feed-back and motor control of artificial limbs. *Sci. Transl. Med.* 6, 257re6–257re6.

[81] Ortmanns, M., Rocke, A., Gehrke, M., and Tiedtke, H. (2007). A 232-channel epiretinal stimulator asic. *IEEE J. Solid State Circuits* 42, 2946–2959.

[82] Paralikar, K., Cong, P., Yizhar, O., Fenno, L. E., Santa, W., Nielsen, C., et al. (2011). An implantable optical stimulation delivery system for actuating an excitable biosubstrate. *IEEE J. Solid State Circuits* 46, 321–332.

[83] Paraskevopoulou, S. E., Barsakcioglu, D. Y., Saberi, M. R., Eftekhar, A., and Constandinou, T. G. (2013). Feature extraction using first and second derivative extrema (fsde) for real-time and hardware-efficient spike sorting. *J. Neurosci. Methods* 215, 29–37.

[84] Peterson, E. J., and Tyler, D. J. (2014). . Motor neuron activation in peripheral nerves using infrared neural stimulation. *J. Neural Eng.* 11:016001.

[85] Polikov, V. S., Tresco, P. A., and Reichert, W. M. (2005). Response of brain tissue to chronically implanted neural electrodes. *J. Neurosci. Methods* 148, 1–18.

[86] Quian Quiroga, R., Nadasdy, Z., and Ben-Shaul, Y. (2004). Unsupervised spike detection and sorting with wavelets and superparamagnetic clustering. *Neural Comput.* 16, 1661–1687.

[87] Rapeaux, A., Nikolic, K., Williams, I., Eftekhar, A., and Constandinou, T. G. (2015). "Fiber size-selective stimulation using action potential filtering for a peripheral nerve interface: A simulation study," in *Proceedings of the 2015 37th Annual International Conference of the IEEE Engineering in Medicine and Biology Society (EMBC)*, Milan, 3411–3414.

[88] Rhew, H. G., Jeong, J., Fredenburg, J. A., Dodani, S., Patil, P. G., and Flynn, M. P. (2014). A fully self-contained logarithmic closed-loop deep brain stimulation soc with wireless telemetry and wireless power management. *IEEE J. Solid State Circuits* 49, 2213–2227.

[89] Richter, C.-P., Matic, A. I., Wells, J. D., Jansen, E. D., and Walsh, J. T. (2011). Neural stimulation with optical radiation. *Laser Photon. Rev.* 5, 68–80.

[90] Robinson, J. T., Jorgolli, M., and Park, H. (2012). Nanowire electrodes for high-density stimulation and measurement of neural circuits. *Front. Neural Circuits* 7, 38–38.

[91] Rosin, B., Slovik, M., Mitelman, R., Rivlin-Etzion, M., Haber, S. N., Israel, Z., et al. (2011). Closed-loop deep brain stimulation is superior in ameliorating parkinsonism. *Neuron* 72, 370–384.

[92] Rossant, C., Kadir, S. N., Goodman, D. F. M., Schulman, J., Hunter, M. L. D., Saleem, A. B., et al. (2016). Spike sorting for large, dense electrode arrays. *Nat. Neurosci.* 19, 634-641.

[93] Rothermel, A., Liu, L., Aryan, N. P., Fischer, M., Wuen-schmann, J., Kibbel, S., and Harscher, A. (2009). A cmos chip with active pixel array and specific test features for subretinal implantation. *IEEE J. Solid State Circuits* 44, 290–300.

[94] Sawan, M., Duval, F., Pourmehdi, S., and Mouine, J. (1990). "A new multichannel bladder stimulator," in *Proceedings of Third Annual IEEE Symposium on Computer-Based Medical Systems, 1990*, Chapel Hill, NC, 190–196.

[95] Schaeffer, L., Nagy, Z., Kincses, Z., Voeroeshazi, Z., Fiath, R., Ulbert, I., et al. (2016). Fpga-based clustering of multi-channel neural spike trains," in *Proceedings of the CNNA 2016; 15th International Workshop on Cellular Nanoscale Networks and Their Applications*, Dresden.

[96] Scholten, K., and Meng, E. (2015). Materials for microfabricated implantable devices: a review. *Lab Chip* 15, 4256–4272.

[97] Scholvin, J., Kinney, J. P., Bernstein, J. G., Moore-Kochlacs, C., Kopell, N., Fonstad, C. G., et al. (2016). Boyden. Close-packed silicon microelectrodes for scalable spatially oversam-pled neural recording. *IEEE Trans. Biomed. Eng.* 63, 120–130.

[98] Schuettler, M., and Stieglitz, T. (2013). "Microassembly and micropackaging of implantable systems," in *Implantable Sensor Systems for Medical Applications*, eds A. Inmann and D. Hodgins (Sawston: Woodhead Publishing), 108.

[99] Seo, D., Carmena, J. M., Rabaey, J. M., Alon, E., and Maharbiz, M. M. (2013). Neural dust: An ultrasonic, low power solution for chronic brain–machine interfaces. arXiv:1307.2196.

[100] Shokur, S., Li, Z., Lebedev, M. A., and Nicolelis, M. A. (2013). A brain–machine interface enables bimanual arm movements in monkeys. *Sci. Transl. Med.* 5, 1946–6234.

[101] Sit, J.-J., and Sarpeshkar, R. (2007). A low-power blocking-capacitor-free charge-balanced electrode-stimulator chip with less than 6 na dc error for 1-ma full-scale stimulation. *IEEE Trans. Biomed. Circuits Syst.* 1, 172–183.

[102] Sooksood, K., Stieglitz, T., and Ortmanns, M. (2009). An experimental study on passive charge balancing. *Adv. Radio Sci.* 7, 197–200.

[103] Sooksood, K., Stieglitz, T., and Ortmanns, M. (2010). An active approach for charge balancing in functional electrical stimulation. *IEEE Trans. Biomed. Circuits Syst.* 4, 162–170.

[104] Stieglitz, T. (2010). Manufacturing, assembling and packaging of miniaturized neural implants. *Microsyst. Technol.* 16, 723–734.

[105] Suaning, G. J., and Lovell, N. H. (2001). Cmos neurostimulation asic with 100 channels, scaleable output, and bidirectional radio-frequency telemetry. *IEEE Trans. Biomed. Eng.* 48, 248–260.

[106] Sun, F. T., and Morrell, M. J. (2014). Closed-loop neurostimulation: the clinical experience. *Neurotherapeutics* 11, 553–563.

[107] Tanghe, S. J., and Wise, K. D. (1992). A 16-channel cmos neural stimulating array. *IEEE J. Solid State Circuits* 27, 1819–1825.

[108] Teo, A. J. T., Mishra, A., Park, I., Kim, Y.-J., Park, W.-T., and Yoon, Y.-J. (2016). Polymeric biomaterials for medical implants and devices. *ACS Biomater. Sci. Eng.* 2, 454–472.

[109] Tronnier, V. M., and Rasche, D. (2015). "Deep brain stimulation," in *Textbook of Neuromodulation*, eds D. Rasche and H. Knotkova (Berlin: Springer), 61–72.

[110] Tufail, Y., Matyushov, A., Baldwin, N., Tauchmann, M. L., Georges, J., Yoshihiro, A., et al. (2010). Transcranial pulsed ultrasound stimulates intact brain circuits. *Neuron* 66, 681–94.

[111] van Dongen, M. N., and Serdijn, W. A. (2016). Does a coupling capacitor enhance the charge balance during neural stimulation? an empirical study. *Med. Biol. Eng. Comput.* 54, 93–101.

[112] Vigraham, B., Kuppambatti, J., and Kinget, P. R. (2014). Switched-mode operational amplifiers and their application to continuous-time filters in nanoscale cmos. *IEEE J. Solid State Circuits* 49, 2758–2772.

[113] Viswam, V., Dragas, J., Shadmani, A., Chen, Y., Stettler, A., Hierlemann, A., et al. (2016). "22.8 multi-functional microelectrode array system featuring 59,760 electrodes, 2048 electrophysiology channels, impedance and neurotransmit-ter measurement units," in *Proceedings of the 2016 IEEE International Solid-State Circuits Conference (ISSCC)*, San Francisco, CA, 394–396.

[114] Ware, T., Simon, D., Rennaker, R. L., and Voit, W. (2013). Smart polymers for neural interfaces. *Polym. Rev.* 53, 108–129.

[115] Wells, J., Kao, C., Konrad, P., Milner, T., Kim, J., Mahadevan-Jansen, A., and Duco Jansen, E. (2007). Biophysical mechanisms of transient optical stimulation of peripheral nerve. *Biophys. J.* 93, 2567–2580.

[116] Wenger, N., Moraud, E. M., Raspopovic, S., Bonizzato, M., DiGiovanna, J., Musienko, P., et al. (2014). Closed-loop neuromodulation of spinal sensorimotor circuits controls refined locomotion after complete spinal cord injury. *Sci. Transl. Med.* 6, 255ra133.

[117] Williams, D. F. (2008). On the mechanisms of biocompatibility. *Biomaterials* 29, 2941–2953.

[118] Williams, I., and Constandinou, T. G. (2013). An energy-efficient, dynamic voltage scaling neural stimulator for a proprioceptive prosthesis. *IEEE Trans. BioCAS* 7, 129–139.

[119] Williams, I., Luan, S., Jackson, A., and Constandinou, T. G. (2015). "Live demonstration: A scalable 32-channel neural recording and real-time fpga based spike sorting system," in *Proceedings of the*

Biomedical Circuits and Systems Conference (BioCAS), 2015 IEEE, Atlanta, GA, 1–5.

[120] Wolf, P. D. (2008). *Thermal Considerations for the Design of an Implanted Cortical Brain–Machine Interface (BMI).* Boca Raton, FL: CRC Press.

[121] Yoo, J., Yan, L., El-Damak, D., Altaf, M. A. B., Shoeb, A. H., and Chandrakasan, A. P. (2013). An 8-channel scalable EEG acquisition soc with patient-specific seizure classification and recording processor. *IEEE J. Solid State Circuits* 48, 214–228.

[122] York, T., Powell, S. B., Gao, S., Kahan, L., Charanya, T., Saha, D., W., et al. (2014). Bioinspired polarization imaging sensors: From circuits and optics to signal processing algorithms and biomedical applications. *Proc. IEEE* 102, 1450–1469.

[123] Zhang, X., Zhang, Z., Li, Y., Liu, C., Guo, Y. X., and Lian, Y. (2016). A 2.89µw dry-electrode enabled clockless wireless ecg soc for wearable applications. *IEEE J. Solid State Circuits* 51, 2287–2298.

Index

About the Editor

Peng Cong received the Ph.D. degree from the Department of Electrical Engineering and Computer Science at Case Western Reserve University in 2008.

He is currently an Engineering Manager and Technology Lead at Verily Life Science (formerly Google Life Science, Google [X]), where he is responsible for hardware R&D (including electronics and mechanical) for implantable devices. Dr. Cong has proven track records in developing and translating technologies for medical devices. He specializes in sensor interface, analog integrated circuit design, microsystem, as well as system integration for wearable and implantable medical devices. Before Google, he was with Medtronic Neuromodulation Core Technology as a Principal IC Design Engineer and leading the Next General Neuromodulation Platform from 2009 to 2014, where he received Technical Contributor of the Year award at Medtronic Inc.

Dr. Cong was a European Solid-State Circuits Conference Subcommittee Member from 2014 to 2016 and has been serving as IEEE Solid-State Circuits Conference (ISSCC) Subcommittee Member since 2015. He also served in committee of three study sections in NIH (Nation Institute of Health) to review Brain Initiative proposals since 2014. He organized and chaired a special section on "Brain-Machine Interfaces: Integrated Circuits Talking to Neurons" and gave a tutorial lecture, "Circuit Design Considerations for Implantable Devices," at 2015 and 2016 ISSCC, respectively. He is currently a Guest Editor for *Journal of Solid-State Circuit*. He also serves on Industry Advisory Board for the Biomedical Engineering Department at UCLA.